The Emergence of Ecological Modernisation

This important new study traces the emergence of the concept of *ecological modernisation*, the attempt to reconceptualise the relationship between the environment and economic development. In the late 1980s it was first argued that it would be possible to modernise industry along ecological lines, creating a springboard to a different kind of economic growth and moving away from the perceived dichotomy between jobs and growth on the one hand, and environmental protection on the other. In this book, a range of international contributions examine the evolution and the implications of this idea.

The Emergence of Ecological Modernisation offers a wealth of empirical research material from across a range of EU and OECD countries, bringing together previously scattered sources for the first time. It addresses a series of theoretical issues that are of key contemporary relevance, such as the relationship between ecological modernisation and sustainable development; strategies for promoting ecological modernisation, and the extent to which it is possible to 'green' contemporary capitalism.

This book represents essential reading for any student of environmental issues, for the economist or political scientist looking for a fresh theoretical perspective, and for practitioners involved in environmental policy-making.

Stephen C. Young is a senior lecturer at Manchester University. He is the author of *The Politics of the Environment* and joint editor of *Environmental Politics and The Politics of Sustainable Development*. He has also contributed to an EU Concerted Action project on Local Agenda 21, and to a comparative, nine-country project on implementing sustainable development in high consumption societies.

Routledge Research in Environmental Politics

1. The Emergence of Ecological Modernisation
Integrating the Environment and the Economy?
Edited by Stephen C. Young

The Emergence of Ecological Modernisation

Integrating the Environment
and the Economy?

**Edited by
Stephen C. Young**

London and New York

First published 2000
by Routledge
11 New Fetter Lane, London EC4P 4EE

Simultaneously published in the USA and Canada
by Routledge
29 West 35th Street, New York, NY 10001

Routledge is an imprint of the Taylor & Francis Group

Typeset in Baskerville by Deerpark Publishing Services Ltd.
Printed and bound in Great Britain by MPG Books Ltd, Bodmin

British Library Cataloguing in Publication Data
A catalogue record for this book is available from the British Library

Library of Congress Cataloging in Publication Data
Young, Stephen C.
    The emergence of ecological modernisation : integrating the
environment and the economy? / Stephen C. Young.
        p. cm.
    Includes bibliographical references and index.
    ISBN 0-415-14173-7 (alk. paper)
        1. Sustainable development–European Union countries–Case studies. 2.
Environmental economics–European Union countries–Case studies. I. Title.

HC240.9.E5 Y68 2000
333.7–dc21   00-057635

ISBN 0-415-14173-7

Contents

Figures

Tables

Contributors

Mikael Skou Andersen is a Research Professor in the Policy Analysis Department of the National Environment Research Institute in Denmark.

Manfred Binder is a Research Fellow at the Environmental Policy Research Unit at the Department of Political and Social Sciences at the Free University of Berlin.

Neil Carter is a Senior Lecturer in Politics at York University, UK, and joint editor of *Environmental Politics*.

Peter Christoff is a Lecturer in Geography at Melbourne University, Australia.

Robyn Eckersley is a Senior Lecturer in the School of Political and Social Inquiry at Monash University, Australia.

Detlef Jahn is Professor of Comparative Politics at the Ernst–Moritz–Arndt University of Griefswald, Germany.

Martin Jänicke is a Professor at the Department of Political and Social Sciences at the Free University of Berlin and Director of its Environmental Policy Research Unit (Forschungsstelle für Umweltpolitik, FFU).

Annica Kronsell is a Senior Research Fellow in the Department of Political Science at Lund University, Sweden.

Charles Lees is a Lecturer in International Relations and Politics at the University of Sussex, UK; and co-editor of the Manchester University Press Series 'Issues in German Politics'.

Philip Lowe is Professor of Rural Economy and Director of the Centre for Rural Economy at the University of Newcastle-upon-Tyne, UK.

Harald Mönch is a Research Fellow at the Environmental Policy Research Unit at the Department of Political and Social Sciences at the Free University of Berlin.

Peter Rawcliffe completed his PhD at the University of East Anglia before moving to work at Scottish Natural Heritage, UK.

Benoît Rihoux is a Lecturer in Politics at the Catholic University of Louvain, Belgium.

Stephen C. Young is a Senior Lecturer in Government at the University of Manchester, UK, and joint editor of *Environmental Politics*.

Preface

During the late 1980s and the 1990s, the concept of ecological modernisation emerged into the policy-making arena in the industrialised democracies. It represents an attempt to reconceptualise the relationship between the environment on the one hand, and economic development on the other. In the 1970s environmental protection was seen within industry and within government as a burden, as creating a choice – jobs and growth *or* environmental protection. But during the late 1980s and early 1990s, some leading industrialists and some policy-makers at the national level and within supranational organisations like the EU and the OECD began to argue that it was possible to reconceptualise the relationship between the environment and economic development. They suggested that a new approach to regulation, and modernising industry along ecological lines in response to the environmental challenge, would create an opportunity, a springboard to a different kind of economic growth. The ensuing capital investment projects, policy initiatives and associated debates brought into focus a beguiling idea – that ecological modernisation offered a way of greening contemporary capitalism so it would be possible to have economic growth that was benign in environmental terms.

This book concentrates on the early years of the 1980s and the start of the 1990s in order to analyse the emergence of ecological modernisation and to set out its main features. These are discussed in the Introduction which draws not just from the general literature, but from the detail of the individual chapters. Chapter 1 examines an early attempt to integrate environment and economy – ecotaxes in Belgium in the 1992–5 period. Chapter 2 analyses the roles of different parts of the green movement in Britain in the late 1980s and early 1990s. Chapter 3 looks at how the issues were dealt with around the same time in the context of the EU's Fifth Environmental Action Plan. Chapter 4 links the features of ecological modernisation to pollution control programmes, focusing in particular on the use of economic instruments in relation to water and the shift in approach from the 1970s to the 1980s in Denmark, France, Germany and the Netherlands. Chapter 5 compares different national experiences during the 1970s and 1980s of the moves in specific sectors from remedial, end-of-pipe approaches to pollution control, to preventative strategies. Chapter 6 reviews the changing relationship between economic growth and environmental protection during the 1980s across the OECD countries, concentrating on energy in particular. As a contrast to the European experience, Chapter 7 draws out the limitations of the British response during the late 1980s and early 1990s. Chapter 8 focuses on the Green Party involvement in subnational coalitions in West Germany during the same period. Chapter 9 is a broader comparative piece highlighting the differences between weak and strong approaches to ecological modernisation. Chapter 10 uses four models to analyse the normative and methodological assumptions embedded in different approaches to integrating economic development and environmental protection.

Apart from focusing on the emergence and features of ecological modernisation, the Introduction also discusses three other issues that are raised by the chapters, and by any detailed discussion of the concept. First, there is the relationship between ecological modernisation and sustainable development. Second, there is the issue of neo-corporatism and whether, in the context of promoting ecological modernisation, industry or government should lead. Lastly there is the extent to which it really is possible to green contemporary capitalism.

This book had its genesis at a European Consortium for Political Research workshop, and the contributors are grateful to that organisation for its support. Thanks are also due to Frank Cass for permission to publish Christoff's chapter, which first appeared in a different form in *Environmental Politics*, 5(3); to Ashgate Publishers for permission to publish a slightly different version of Andersen's chapter from that which appeared in M. Joas and A.-S. Hermanson (eds.) (1999). *The Nordic Environments: Comparing Political, Administrative and Policy Aspects*; and to Manchester University Press for permission to use three of the charts in Andersen's chapter that were in his 1995 book, *Governance by Green Taxes*.

Lastly, as editor I would like to thank the contributors for their patience; and Joek Roex for his tremendous work on the copy editing, and in preparing the final manuscript for the publishers.

Stephen C. Young
University of Manchester

Abbreviations

AL	*Alternative Liste* (Berlin's local Green Party)
BATNEEC	Best available technique not entailing excessive costs
CAP	Common Agriculture Policy (of the EU)
CDU	Christian Democratic Union (in Germany)
CFCs	Chlorofluorocarbons
CO_2	Carbon dioxide
CPRE	Council for the Protection of Rural England
CSU	Christian-Social Union (in Germany)
CVP	*Christelijke Volkspartij* (in Belgium)
DoE	Department of the Environment (in UK)
DoT	Department of Transport (in UK)
DG	Directorate General (of the European Commission)
DIW	*Deutsches Institut für Wirtschaftsforshung*
DIY	Do-it-yourself
EEB	European Environment Bureau
EEC	European Economic Community
EIA	Environmental impact assessment
EU	European Union
FCCC	Framework Convention on Climate Change
FDF	*Front Démocratique des Francophones* (in Belgium)
FDP	Free Democratic Party (in Germany)
FFU	Forschungsstelle für Umweltpolitik
FoE	Friends of the Earth
FRG	Federal Republic of Germany
GAL	*Grüne Alternative Liste* (Hamburg's local Green Party)
GATT	General Agreement on Tariffs and Trade
GCC	Global Climate Coalition
GDP	Gross domestic product
GNP	Gross national product
HCFCs	Hydro chlorofluorocarbons
HMIP	Her Majesty's Inspectorate of Pollution (in UK)
ICM	Integrated crop management
IUCN	International Union for the Conservation of Nature

LCA	Life cycle analysis
LDCs	Less developed countries
LDP	Liberal Democratic Party (in Japan)
MEP	Member of the European Parliament
MITI	Ministry of International Trade and Industry (in Japan)
MP	Member of Parliament
NEPP	National Environmental Policy Plan (in the Netherlands)
NGO	Non governmental organisation
NICs	Newly industrialising countries
NIMBY	Not in my back yard
NRA	National Rivers Authority (in England and Wales)
OECD	Organisation for Economic Co-operation and Development
OMMRI	Ontario Multi-Material Recycling Incorporated (in Canada)
PDS	Party for Democratic Socialism (in Germany)
PRL	*Parti Réformateur Liberal* (in Belgium)
PS	*Parti Socialiste* (in Belgium)
PSC	*Parti Social-Chrétian* (in Belgium)
PVC	Poly-vinyl-chloride
PVV	*Partij voor Vrijheid en Vooruitgang* (in Belgium)
RSNC	Royal Society for Nature Conservation
RSPB	Royal Society for the Protection of Birds
RWI	*Rheinisch–Westfälisches Institut*
SEA	Single European Act
SP	*Socialistische Partij* (in Belgium)
SPD	Social Democratic Party (in Germany)
SMEs	Small and medium-sized enterprises
TNCs	Trans national corporations
USA	United States of America
UNCED	United Nations Commission on Environment and Development
VLD	*Vlaamse Liberalen en Democraten* (in Belgium)
VOC	Volatile organic compounds
VU	*Volksunie* (in Belgium)
WCED	World Commission on Environment and Development
WRI	World Resources Institute
WWF	World Wide Fund for Nature

Introduction

The origins and evolving nature of ecological modernisation

Stephen C. Young

During the 1990s, ecological modernisation attracted increasing attention. In the industrialised democracies, it began to dominate as a way of approaching environmental policy. It seemed to represent 'a new consensus on how to conceptualise the environmental problem, its roots, and its solutions' (Hajer 1995: 101; see also Chapter 9). However, the initial problem with discussing ecological modernisation is that 'There is no one canonical statement' (Weale 1992: 75). The result is that writers have approached ecological modernisation from different perspectives. Mol (1995: 27–8) has identified three main ways in which it is interpreted.

First, ecological modernisation is used in the context of overarching debates about social theory. These range from discussions about the evolving relationships between institutions and the environment in environmental sociology (Mol 1995: chap. 2; Spaargaren and Mol 1992), through to broader theorising about modernity and postmodernity (Giddens 1990; Beck *et al.* 1994). Secondly, it is presented as a new paradigm by social scientists analysing the changing nature of environmental politics and policies in the 1980s and 1990s. Weale (1992) is a prominent example of this approach. Thirdly, ecological modernisation is adopted in a prescriptive way to refer to a programme of environmental and economic policies designed to tackle the range of ecological problems facing governments in the industrialised democracies at the end of the twentieth century. This version is often used normatively by political parties and non-governmental organisations(NGOs), as it has been in the Netherlands. Jänicke's early work provides a fourth understanding of the term. He argues that some companies promoted ecological restructuring simply as a cost-cutting approach to enhance competitiveness in the shift from smokestack industry to clean technology (Jänicke, 1985; see also Chapters 5 and 9). Such technological adjustment had little to do with environmental gains. Gouldson and Murphy (1997a) also develop this interpretation.

This lack of a clear defining statement is confounded by the way in which writers adopt the term ecological modernisation without always being clear about the sense in which they are using it. Hajer (1995: 29) argues that the meaning of the term only becomes clear over time: 'A new policy discourse as comprehensive as ecological modernisation is not conceptualised as one united set of ideas, but only gradually emerges after years of institutionalised debate'.

The main focus in this Introduction is on the second body of writing, although the others are also discussed to some extent by the contributors to this book. In essence, this interpretation of ecological modernisation is about reconceptualising the relationship between the environment and the economy in the industrialised democracies (Weale 1992; Mol 1995; Hajer 1995; Dryzek 1997: chap. 8). In the 1970s in the wake of the Club of Rome report (Meadows *et al.* 1972) and the 1972 Stockholm conference, the conventional analysis was that there was a choice. It was either jobs and growth, or environmental protection – not both. It was widely believed that enhanced environmental protection would lead to more constraints on economic growth. But during the 1980s and early 1990s, ecological modernisation increasingly emerged as a theory based on a new line of argument.

The central point had become the following: it was possible to modernise industry along ecological lines in response both to the developing environmental challenge; and to the resulting pressures from government and civil society. The core argument here was that ecological modernisation could lead to *a qualitatively different kind of economic growth.* Growth could be maintained within a framework of stronger environmental protection, making the growth more benign in environmental terms than traditional patterns of economic growth had been. These had seen the environment as a sink for wastes, and resources as free goods for industry to exploit. But ecological modernisation understands the importance of protecting the environment, because business ultimately depends on the health of the planet and surrounding atmosphere. Industry's future prosperity depends on what Mol (1995) calls its 'sustenance base'. It is thus in industry's interest to develop cleaner technology and act in different ways in order to protect the planet's capacity to support human life, and provide resources so that the sustenance base itself is protected. According to this analysis environmental protection ceases to be an extra burden as it had been seen in the 1970s. It becomes an opportunity – a springboard to a different kind of economic development. It becomes possible to adapt traditional treadmill capitalism, with its cavalier disregard for the environmental consequences of economic growth, in the light of the developing environmental challenge. The beguiling vision that emerges is a 'win–win' scenario in which it is possible to have higher living standards and other benefits of economic growth, together with enhanced environmental protection. Several writers refer to this as uncoupling economic growth from increasing environmental stress. As a result environmental protection and economic prosperity 'are seen as properly proceeding hand-in-hand' (Dryzek 1997: 145; see also Chapters 4 and 6).

This Introduction leads into a book that focuses mainly on ecological modernisation in the sense of the emergence of this new paradigm. The Introduction discusses the changing nature of environmental politics and policies in the 1980s and 1990s, and provides links with the remaining chapters. The first section sets out eight features of ecological modernisation. Subsequent sections draw from the detail presented. The second section summarises the reasons for the emergence of ecological modernisation, and identifies the factors driving it forward.

The third assesses its importance by the late 1990s, and its relationship to sustainable development. The next two sections analyse its evolving nature and future prospects; and the debates about whether industry or government should take the lead in promoting it. The discussion reveals gaps in our understanding of the scope and nature of ecological modernisation as interpreted here. The last section identifies the aspects of ecological modernisation that appear to have the most interesting potential for further research. The main purpose of this book is to analyse the emergence of ecological modernisation. Beyond that, it aims to contribute to the debates about how far it is possible to green contemporary capitalism; and about the extent to which ecological modernisation is compatible with sustainable development.

Features of ecological modernisation

Companies adopting a longer-term perspective

During the 1980s, many big companies began to adapt their approach to medium and long-term planning in the light of the developing debates about environmental issues. They revised their views about the nature and extent of their responsibilities in terms of the environment. The first feature of ecological modernisation is the way in which companies have shifted from simple compliance over pollution regulations to a much broader interpretation of their environmental responsibilities. This approach of putting environmental issues at the heart of a company's corporate planning emerged first in America (Fischer and Schot 1993; Schmidheiny 1992). It was initially promoted by companies like Minnesota Mining and Manufacturing (Tibbs 1993). Both before but especially after the 1987 Brundtland Report (WCED 1987), it spread to Europe. The subsequent establishment of the Business Council for Sustainable Development, the 1992 Rio Earth Summit and the subsequent conferences on climate change and other topics, led to further interest (Willums and Goluke, 1992).

Discussions about ecological modernisation usually focus around clean technology and the elimination of polluting emissions to air and water from manufacturing plants. However the range of impacts that companies make on the environment is much more complex than this. Wood (1995) shows this by setting out the five stages of a product's life cycle – from the supply of raw materials and the production process, through the retail stage and the way the consumer uses the product, to the post-consumer stage. The latter relates to the product's impacts on the environment after it has been discarded – whether the constituent parts remain in the environment, as toxic chemicals for example; go to incineration or landfill; or are reused or recycled in some way. By the 1990s, new patterns of behaviour were emerging at each of these five stages of a product's life cycle.

First there are the cases of firms establishing a new approach towards their suppliers. In the mid-1990s, Unilever replaced fish oil in fish products with soya or palm oil. Its source of fish oil had been the industrial fishing companies, whose boats hoovered up the small fish at the bottom of the food chain. These are a food source for bigger fish, as well as for birds and seals, and thus help to maintain fish stocks. Unilever was one of the largest purchasers of fish in the world, and had become concerned about the long-term maintenance of fish stocks (*Environmental Digest* 1996, 4: 3; *The Independent* 23 April 1996; *Green Futures* 1997, 4: 37). The case of timber companies supplying pulp to paper users shows how user pressures can build up over time and fundamentally change the suppliers' approach. Macmillan Bloedel went from strongly defending the practice of clearing pristine rain forest in British Columbia in 1994, to accepting that such practices were unsustainable and not viable in economic terms 4 years later. This was because companies like Scott, one of the largest UK manufacturers of toilet rolls and paper products, and the *New York Times* cancelled big pulp contracts (*Greenpeace Business* April 1994: 1 and June 1998: 1).

At the second stage – the production process – there are numerous examples of firms developing new kinds of capital investment programmes in order to promote clean technology and reduce, or even eliminate, damage to the environment (Irwin and Hooper 1992; Jackson 1993; Clift and Longley 1995; Porter and van der Linde 1995; Hillary 1997; see also Chapters 4 and 5). In the 1970s and 1980s, the focus was on the routine replacement of equipment; the incorporation of new technology irrespective of environmental considerations; and on the installation of robots and other automated processes that saved labour costs. But by the late 1980s and early 1990s, strategic planners and corporate research departments were developing ways of adapting production processes to reduce or eliminate harmful environmental effects. In promoting the ideas of eco-efficiency and industrial ecology (Marstrander 1994), they sought to redesign processes so as to remove the source of each harmful environmental effect. This approach saves energy, conserves resources and reduces costs while still adding value. Bleach-free paper production is often cited. Vehicle manufacturers were more ambitious. They moved, albeit with varying degrees of speed, towards including recycled materials, recycling platinum, incorporating reusable parts and recycling the quantities of water used during the production process (ETA 1995; Schnaiberg 1997). In addition, some paint lines now operate as closed systems. This makes it possible to collect the solvent vapours – the volatile organic compounds (VOCs) – as the paint dries, and either incinerate them safely, or reuse them in liquid form when making up new paint. This avoids polluting the neighbourhood with paint fumes and prevents the VOCs from contributing to atmospheric pollution and summertime urban smog.

At the retail stage several different pressures reinforce each other. Retailers themselves demand new kinds of equipment. Thus supermarket chains are trialing and buying new kinds of refrigeration and air-conditioning equipment based on ammonia, CO_2, hydrocarbon and other CFC-free technologies (*Greenpeace Business* February 1998: 7). Where big brewery chains take the initiative and install such equipment in the bars and retail outlets that they supply, the pressures are reversed. In such cases the retailers have to improve their standards. In addition retailers have been greatly affected by the EU directive requiring companies to recover half of their waste (*Environmental Digest* 1996, 6: 19–20).

Retailers deal with packaging both when receiving goods and when selling and delivering them.

In addition, retailers sometimes force their own suppliers to reduce their impacts on the environment. Paint retailers, for example, have pressurised manufacturers to produce paints containing smaller percentages of VOCs. The World Wide Fund for Nature (WWF) set up 'the 1995 Group' in 1989, with the aim of getting timber users to phase out uncertified suppliers by 1995. The users were prepared to commit themselves to dealing only in timber that could be shown to have come from sources certified as being managed in environmentally sensitive ways (Wood 1995: 13–5). DIY stores, manufacturers of furniture and garden products, and furniture and more general retailers joined the timber merchants as members. The Group uses the standards of the Forest Stewardship Council when deciding whether suppliers are using sustainable forestry management techniques. To remain as members, firms have to show that they are steadily increasing the proportion of their supplies coming from certified sources. This reinforces both members taking on responsibility for the environmental impacts of their suppliers, and the pressure on suppliers to adapt their routine approaches. Retailers and timber users are becoming involved in managing the basic resource.

The fourth stage in the product life cycle is producers trying to influence the way purchasers actually use products. Bayer and other chemical companies belonging to the European Crop Protection Association now take greater responsibility than in the past for the ways in which farmers use the fertilisers and pesticides they supply (Mol 1995: chap. 8; Wood 1995). Initiatives seem to have emerged first in Germany in the late 1980s, and then spread to Britain and other parts of Europe. Within the chemical companies themselves, support grew for the view that continuing reliance by farmers on chemicals would damage soil fertility in the long term and would not be sustainable. Instead there was a need to develop specialist advisory services to promote integrated crop management (ICM). The aim is to reduce farmers' dependency on chemicals, to encourage crop rotation and the use of disease resistant varieties, and mechanical and biological control of pests. Demonstration farms were also established to show practical results as part of a European-wide network of ICM projects. Between 1991 and 1995, 450 British companies joined the Linking Environment and Farming part of this network.

At the post-consumer waste stage companies are starting to respond to pressure, and accepting responsibility for the impact of their goods on the wider environment after consumers have finished with them. A good example is the Ontario Multi-Material Recycling Incorporated (OMMRI) case in Canada. OMMRI was created by firms from six sectors – grocery products manufacturers, grocery products distributors, packaging manufacturers, producers and processors of plastics, suppliers and users of printing paper and manufacturers of soft drinks. When it started it concentrated on promoting recycling, using kerbside blue boxes. The amount handled rose from 150,000 tonnes in 1988 to 450,000 tonnes in 1994. Thereafter, OMMRI began to focus more on sharing its experience more widely and on R&D in relation to the development of new markets for recycled materials – as with making mountain cagoules out of melted-down plastic bottles. Other examples of companies accepting responsibility for post-consumer waste on a voluntary basis are telephone companies organising projects to recycle old phone directories, and oil companies promoting local schemes to collect waste oil for recycling and safe disposal. This prevents oil from being put down drains, and getting untreated into water courses. In some cases, consumer pressure forces companies to change the materials they use. Two of the largest Spanish producers of bottled water stopped using polyvinyl chloride (PVC) bottles because of boycotts. The chlorine content of PVC means recycled plastic waste streams become contaminated and, if incinerated, are a source of toxins. The companies used different containers not just because of energy costs at the production stage, but because of the post-consumer waste they were generating (*Environmental Digest* 1996, 9: 13; *The Guardian* 20 September 1996). In other cases, governments have taken the initiative and converted voluntary schemes into mandatory ones. Examples here include the 1996 decisions in Denmark to recover 75 per cent of nickel-cadmium batteries, and in Germany to make the final owner of a car responsible for delivering it to approved recycling centres (*Environmental Digest* 1996, 8: 12 and 1996, 11: 16).

Most writers focus on manufacturing industry (see Chapter 5). But this trend of companies developing a broader view of their environmental responsibilities is also in evidence in the service sector where the five-stage product life cycle does not fit. The banks for example deal with all kinds of firms: suppliers, producers, distributors and retailers. By imposing particular conditions, they can influence the way in which their customers approach their environmental responsibilities. In 1991 the United Nations Environment Programme began to work with banks to produce a *Statement by Banks on Sustainable Development.* The six initial signatories of the 1992 Statement swelled to seventy by 1995 (Wood 1995: 27–8). The Statement emphasised the importance of the precautionary approach; the need for companies to include environmental risk in their risk assessment procedures; and the need for firms not just to comply with relevant regulations at all levels from the local to the international, but to develop sound environmental management practices. These do not represent the application of strict new conditions being attached to funding arrangements. But the NatWest Group, for example, developed mechanisms to change customers' perceptions of how they interact with the environment. After signing the Statement, it trained its lending managers to assess environmental risk with regard to commercial loans; and produced a software model to help its customer companies to identify ways in which they were having an impact on the environment. Consultants have also exploited this shift in approach. Their activities have expanded considerably into the sphere of ecological modernisation.

The insurance sector provides a more extensive example of non-manufacturing companies changing their ideas about their environmental responsibilities. In the past, insurance was simply a matter of collecting lots of premiums in order to pay out on occasional claims. But several factors have undermined this strategy, collectively threatening the industry's viability (Salt 1997; Salt, 1998; *Economist* 28 February 1998: 93–5). Climate change is believed to be contributing to the growing numbers of hurricanes, droughts, forest fires, floods from rivers and rising sea levels and other environmental catastrophes. It is the increased frequency of such events that has alarmed the industry. The 1992 Hurricane Andrew in Florida and the 1994 Los Angeles earthquake bankrupted a number of property insurers because the second disaster occurred so soon after the first. The future viability of the industry depends on the world's capital markets. They are the only place on the planet with enough assets to cover both large scale, and more frequent, catastrophes. Insurance companies control up to 10 per cent of the world's financial flows via their activity in equity markets, trying to ensure that their investments grow sufficiently to cover increasing insurance claims. So, for a number of reasons, the insurance industry has a vested interest in helping to tackle the problem of climate change.

In 1996 the United Nations Environment Programme built on movements within the industry in establishing the Insurance Industry Initiative. The aim is to get insurance companies involved in relationships with other organisations that help them both to foresee environmental catastrophes and to reduce their impact. This includes pressurising their clients and commissioning research to improve their understanding of environmental change. They have also lobbied governments and other actors to persuade them to stop procrastinating and take decisive action on climate change, to raise construction standards to make buildings less vulnerable to windstorms and to promote technology transfer. The industry's future viability depends on its ability to improve its relationships with, and to increase its influence over, governments and clients on these issues. Developers and industrialists are offered premium discounts where action to reduce risk has been taken (Freeman and Kunreuther 1997). It is not managing a resource like timber differently or changing production processes. But it is responding to environmental change and starting to operate in equivalent ways in order to protect the industry's long-term viability.

The shift in gear in industry's approach from taking the environment for granted, to accepting a wider responsibility for environmental issues, is also borne out by the response of trade associations and other sectoral interest groups. The examples of the European Crop Protection Association and the Forestry Stewardship Council have been mentioned above. The latter has a fishing equivalent called the Marine Stewardship Council, established in 1996, and similarly promoted by WWF (*Environment Digest* 1996, 2: 2; *Green Futures* 1997, 4: 37). Sometimes the initiative comes from within the industry. In the US the Association of Post-Consumer Plastics Recyclers actively promoted a shift away from PVC plastics. In addition, Mol (1995) has shown how, in the case of Dutch chemicals, such industry groups began to argue the importance of taking a longer-term perspective. Others have highlighted the same point in the analysis of lobbying activity in Brussels (Mazey and Richardson 1993; see also Chapter 3).

What emerges from the discussion so far is that, in a variety of sectors, including non-manufacturing ones, industrial leaders have come to the view that they can only achieve long-term economic success if they take on wider responsibilities to the environment and adapt their behaviour accordingly. This involves conserving resources more effectively and not treating the environment simply as a sink for wastes. The argument goes that we ignore the impact of economic activity on natural capital when using conventional measures of profit, growth and gross national product. The point is well made by Gray's summary of Daly's argument:

> Daly's point [...] is the commonly accepted notion in economics, business and accounting that prudent behaviour suggests we only take as income that which is left over after maintaining our capital intact [...] What we currently measure as 'income' does *not* leave our *natural* capital intact – it leaves it depleted. It must follow, therefore, that our measure of income is wrong and the level of consumption that we have enjoyed has been paid out of capital.
>
> (Gray 1996: 184, author's emphasis)

New corporate strategies

The second feature of ecological modernisation is the direct consequence of the first. If industry is to deliver on its acceptance of wider environmental responsibilities, then managers need to change the ways companies are run and develop new strategies. The focus here is inside the company, on establishing new attitudes, operationalising new ideas, adapting management techniques and developing new tools (Beaumont *et al.* 1993; Fischer and Schot 1993; Welford 1994). Some writers have dealt with this in a rather general way (Elkington 1987; Hunt and Johnson 1995). Others, however – Welford (1996); Marstrander (1994) for example – have argued that there is a substantial difference between adapting businesses to environmental concerns and managing companies so as to control and then reduce their environmental impacts. These represent two distinct strategies. They correspond with Hajer's distinction between remedial, cleaning-up approaches and anticipatory or preventative ones (1995: 34–5; see also Chapters 4, 5 and 9). The first is concerned simply with responding in a limited way to regulatory changes introduced by government and to public concerns about the environment. The second strategy is much more ambitious. It is about changing the way the firm is run over time so as to reduce – continuously – its overall impact on the environment.

The first strategy links clearly to the first two steps on Wood's 'Ladder of Progress' diagram (Wood 1995: 10). Step 1 simply focuses on compliance, on firms meeting regulations on solid waste and emissions to air, land and water. Step 2 concerns compliance and improving company performance in areas not subject to regulation – as in energy and transport. This relates to projects that bring financial gains. This limited response is carried out through a mix of environmental charters, end-of-pipe technology, minimising waste, reducing energy and new environmental standards. This first strategy sets managers a new agenda and leads to the establishment of environmental monitoring systems to identify operational problems.

In contrast, the second strategy is preventative, not remedial. It aims at continuous reduction in a firms' environmental impacts in both the short and the long term. Environmental management systems aim to identify areas where the firm's environmental performance can be improved. This represents a much more focused approach than is evident with the remedial strategy. It is carried out by incorporating clean technologies into capital investment programmes, and developing risk assessment techniques and more ambitious approaches to environmental impact assessment (EIA) and life cycle analysis (LCA). The purpose of LCA is to evaluate 'the total environmental load associated with providing a service, by following the associated material and energy flows from their 'cradle' (i.e. primary resources) to their 'grave' (their ultimate resting place, as with solid waste or dispersed emissions' (Clift and Longley 1995: 111). This technique has been used since the late 1980s by multinational corporations and large European companies in chemicals, plastics, electricals, vehicles production, paper, food, building materials, aluminium, consumer products and packaging (Berkout 1996). Ford used LCA to investigate alternative paint systems with BP Chemicals. Firms like Volvo and BMW have used LCA to analyse whether replacing steel with aluminium in vehicle production, and recycling more parts, transfers the environmental load upstream to suppliers or downstream to users – or leads to a reduced environmental load in overall terms.

The second strategy, the preventative, continuous reduction one, links to the third, fourth and fifth steps on Wood's ladder: environmental management that aims at continuous improvement, producer responsibility for the environmental impacts of suppliers and products, and engagement with a wider range of stakeholders than before. It reveals the way that part of ecological modernisation is industry accepting that it has to operate within tighter environmental limits. Ultimately, the commitment in this second strategy to a continuous reduction in environmental impacts opens up a much wider agenda for a company, involving the need for 'radical change in its corporate culture' (Welford and Starkey 1996: 172).

In theory, the second strategy's aim of continuous reduction would make it possible for a company to reach the ideal model at the top of Wood's ladder. This is the sustainable company operating within the biosphere's limits. It is difficult in the mid-1990s to envisage exactly what this might involve. The idea of a company having, in total, positive impacts on the environment can be linked to cases like a small firm that is managing woodland for timber and wildlife, earning income by selling its products locally, from green tourism initiatives like cycle hire, and by providing accommodation for tourists, using local food sources to feed them. But it seems a paradoxical concept when applied to a multinational.

However, it is possible at the end of the twentieth century to identify a two-stage process starting to emerge from the continuous reduction of environmental impacts strategy. To begin with, companies make substantial changes to their products in the light of environmental considerations. Later, the changes start to affect their core business activities, and they start to evolve into a different species of firm. The mid-1990s provide lots of examples of the first part of this strategy – significantly adapting products. Hoover and Lucas, respectively, won green awards for producing environment-friendly washing machines and a new fuel-injection system. The car makers began to move on from petrol-driven vehicles to focus on solving the problems presented by fuel cell technologies – as with electric cars and buses. Electrolux, the Swedish-based fridges and freezers company, took environmental considerations more seriously, systematically redesigning its entire product range. It extended the use of the non-CFC producing hydrocarbon technologies, and improved the energy efficiency of its products. In addition it started to design for recycling so that materials that could not be recycled and re-used were eliminated as far as possible, thus closing the materials loop.

By the late 1990s, signs of firms shifting from significantly adapting products to changing the nature of the company were emerging. The oil sector provides a good example of how, because of a longer-term perspective, the core of the business begins to change. Shell and BP began to invest in offshore wind energy and solar heating systems. In 1997 Shell even joined the European Wind Energy Association! Although investing in producing renewable energy equipment can be interpreted as a precaution or token gesture, it also opens up the possibility of the oil giants of the late twentieth century changing their core businesses to become the world energy leaders of the twenty-first century. Similarly, it is possible to envisage the car companies opportunistically moving on from producing relatively pollution-free cars to building electric buses and becoming extensively involved in public transport – as Fiat are beginning to do. The extent to which companies are moving on from the stage when they develop environment-friendly products to change their core business is as yet limited. But examples like these do show how companies can reduce their long-term environmental impacts and operate much more within the biosphere's limits – whilst simultaneously maintaining profitability. It would be misleading to interpret this as merely fitting into the traditional pattern of companies exploiting diversification opportunities. It is about the emergence of a new species of company which accepts that it has to operate within tighter environmental limits than before. For Stead and Stead (1996), if firms are to survive in the long term, it is critical that sustainability becomes a core value. For Welford (1996: 271), the techniques underlying the continuous reduction strategy 'challenge the very way we do business'.

Integrating environmental and economic policies within government

The next feature parallels industry's experience on the government side. Just as industry had moved away from end-of-pipe treatments, so governments began to shift away from cosmetic approaches to the environment. Instead of remaining an add-on, the environment was moved nearer to the heart of government decision-making processes. Increasingly decision-makers came to accept the need to integrate economic and industrial policies with environmental policies, whilst pushing ahead with strategies to modernise economies and state institutions (Cairncross 1995). The aim here was to inject environmental considerations into policy-making processes on issues like industry, energy, transport and trade, so that, where possible, environmental problems were removed at source.

The focus on integrating economic and environmental policies represented a new approach. In the 1960s and early 1970s, the emphasis of economic and industrial policy-making had been on growth and jobs (Hayward and Watson 1975). In addition, as interest in ecological modernisation began to grow, economic and industrial policies were integrated with programmes on urban development, land-use planning, transport and other aspects of infrastructure planning. This was all part of the wider, integrating approach to reduce energy and resource intensive economic activity (Owens 1997). Gouldson and Murphy (1997b: 82) argue that the promotion of ecological modernisation depended on the redefinition of 'the boundaries, objectives, functions, and cultures of government institutions concerned with both environment and economy'.

The initial moves to integrate environment and economy took place first in the Netherlands, West Germany and the European Commission (see Chapters 3–6, 8 and 10). They are symbolised by the early attempts to integrate both the prevention of, and control of, pollution, thus drawing together the economic and industrial dimensions and the environmental. Later chapters show how once governments began to try to integrate environment and economy, they were constrained by the multidimensional complexity of problems. Increased traffic congestion, for example, leads to local residents suffering from noise, toxic lead fumes, and increased asthma rates. New road building damages wildlife sites, spreads development away from built-up areas and public transport routes, generates car use and creates a demand for aggregates which have to be transported to the sites in question. Tackling increased car congestion thus involves not just public transport policy, but health, land-use planning and biodiversity issues. The response in Britain and southern Europe was much slower than in northern Europe (Weale 1992; see also Chapters 1–2 and 7). The moves to integrate what had been separate policy-making streams needs to be distinguished from the frequently made point about environmental issues moving up the political agenda. The latter relates to politicians judging the environment in terms of its electoral salience, in terms of its importance relative to health, education, jobs and so on. Ecological modernisation is about a broader approach to economic and industrial policy-making. Its relationship to the 1987 Brundtland Report and to the moves towards sustainable development are discussed below.

New policy tools

Just as industry's increasing concern with the environment led it to search for new ways of running companies, so governments developed new techniques and instruments to implement policy. In the nineteenth and twentieth centuries, grow-

ing concerns about environmental protection had led to the development of a
regulatory, command-and-control approach. Standards were established, above
which factories could not pollute. However, the experience of the 1970s and early
1980s showed this approach to be slow, legalistic, inflexible and only erratically
effective (OECD 1989; Vig and Kraft 1990; Weale 1992: chap. 1; Wintle and
Reeve 1994; Mol 1995). Part of ecological modernisation has involved finding
ways around this implementation deficit (see Chapters 5 and 6). Environmental
economists developed economic instruments as a more flexible means of encouraging companies to act in ways that minimise environmental damage. A simple
example is landfill taxes which create an incentive for firms to minimise their
wastes and find other uses for it. This is one of the most discussed aspects of
ecological modernisation (Andersen 1994; Cairncross 1995; Eckersley 1995: see
also Chapters 1, 3–5, 7, 9 and 10). The incentive aspects of economic instruments
could be used to establish specific projects. In 1997, Shell and Pilkington combined
in a joint venture to produce solar powered equipment. There was financial backing from the German federal government, and from North Rhine-Westphalia,
with the latter undertaking to buy half the annual output (*Greenpeace Business*
December 1997: 1). This is not industry or government leading. It is both accepting
that each has a role, that the two are dependent on each other.

A number of other techniques were established to help policy-makers take
greater account of the environmental, as distinct from the economic dimensions
of alternative policies and projects (Wathern 1988; O'Riordan and Jordan 1995;
Roberts 1995; Welford and Starkey 1996; Wilson and Bryant 1997). These
include, in particular, the precautionary principle, environmental impact assessment (EIA), best available technique not entailing excessive costs (BATNEEC),
best practical environmental option, and environmental risk assessment. These
techniques were applied to government projects as well as to private sector
proposals.

Partnerships and participation

Initially, part of ecological modernisation was a more open and inclusive
approach to policy-making. Both Weale (1992: 167–180) and Hajer (1995: 28–
9) highlight enhanced participation processes. However, terms like partnership
and participation tend to be generalised, umbrella terms. Different aspects need
to be distinguished here. Although many existing institutions were adapted to
cope with ecological modernisation, new ones also appeared. Examples include
the Forest Stewardship Council at the supranational level; and round tables at the
national and subnational levels, where there were also waste forums and green
business clubs. At a different level, big companies developed arrangements to
draw in the views of their stakeholders – shareholders, lenders, regulators, policy-
makers and customers. Some stakeholders become, in effect, the representatives
of future generations. At the neighbourhood level, around major plants, there
were frequently partnerships with employees, the local community and other
local stakeholders. For example many firms undertook landscaping works

around their plants and around schools and other community buildings. This improved both habitats for wildlife and the quality of life for local people.

This more open and participative approach was further encouraged by the post-Rio process. The emphasis in Chapter 28 of the Rio agreement on drawing minorities in and generating consensus-based approaches was significant here. Local Agenda 21 initiatives added to previously established procedures like the EIA processes and created additional opportunities for participation (Lafferty and Eckerberg 1998). Community participation sometimes led to amendments to proposals at the site level. But, there were lots of problems (Young 1996). The impact of these approaches on policy frameworks was much more mixed.

On the other hand, a more enhanced role certainly emerged for NGOs. In the 1980s some NGOs began to shift away from their confrontational strategy of the late 1960s and 1970s (Rudig 1988). They became 'less radical, more practical and more policy-oriented' (Hajer 1995: 93; see also Chapters 2, 3 and 7). Greenpeace's shift from using direct action and stunts to try to stop the dumping of wastes at sea, to searching for alternative solutions, as with its greenfreeze, CFC-free fridges, is a typical example. Melchett, Greenpeace Chief Executive, summarised this as follows: 'The new environmental struggle is to put solutions into practice. Solutions are only prevented from becoming mainstream because they are suppressed, [...] held back by specific vested interests in business which starve them of investment' (*Financial Times* 26 September 1996). Environmental NGOs thus adopted 'the role of counter-expert, illuminating alternative policy solutions' (Hajer 1995: 94), and challenging 'the assumptions on which traditional policy elites operated' (Weale 1992: 170) with regard to nuclear power and to air, water, soil and marine pollution. They focused at national and supranational levels working with industry, and lobbying governments to amend policy proposals. Knox and Taylor (1995: fig. 1.1) highlight how international NGOs relocated themselves to cities where the European Commission and multinationals were established. Weale (1992: 167–80) and Mol (1995: 378) show how NGOs in Germany, Denmark and the Netherlands got drawn into policy dialogues and taken more seriously. These changing approaches were paralleled at the subnational level.

However individual NGOs were sometimes a bit schizophrenic over this. They continued with stunts, non-violent direct action and media-based approaches, while simultaneously working quietly to produce solutions. The example of talking to users of pulp and paper while campaigning to embarrass logging companies in public has been referred to above. Similarly, Greenpeace campaigned against BP's proposed oilfield west of the Shetlands, while working quietly with the same company on constructive solutions with solar energy. The extent to which the environmental movement was divided varied from country to country. But an important part of ecological modernisation was about adopting more inclusive approaches and marginalising the radicals (Taylor 1991). It is also clear that not all NGOs and environmental movements responded in this way. Rawcliffe, for example, shows how some British anti-roads groups continued to be radical and uncompromising in the 1990s, arguing for deindustrialisation not modernisation (Chapter 2).

During the late 1980s and the early 1990s, a variety of new and adapted partnerships emerged. Most were open policy networks. But a few were more closed policy communities. These arrangements included government organisations, trade associations, individual companies and their stakeholders, and moderate NGOs looking for solutions. There were links to LA21 and other participatory processes. Collectively these partnerships represented changing attitudes to the need to develop new approaches to reconcile economic growth and environmental protection. But their main importance to the emergence of ecological modernisation was their contribution to the development of collaborative frameworks. Dryzek (1997: 144) argues that in Germany, Japan, the Netherlands, Norway and Sweden a pattern of collaborative working was becoming clear in the 1990s: 'Ecological modernisation implies a partnership in which governments, businesses, moderate environmentalists, and scientists co-operate in the restructuring of the capitalist political economy'. Scientists are included as they played a key role.

Science and technology

Implicit in the features of ecological modernisation discussed so far is a stronger, more central role for science and technology (Mol 1995: 24, 39–40, 372–5; Simonis 1989; Cairncross 1995; see also Chapters 4 and 9). In industry, research and development programmes have focused to an increasing extent on using science and technology to reduce and eliminate industry's environmental impacts. This is symbolised by the shift from the remedial strategy involving end-of-pipe technology to the preventative, continuous reduction strategy based around clean technology processes in the 1980s and the 1990s. The use of LCA is a typical example (Irwin and Hooper 1992; Jackson 1993; Clift and Longley, 1995).

NGOs have also developed stronger links to scientists as part of their campaigning to pressurise firms into adapting new techniques. After the Montreal Protocol, some chemical companies planned to use environmentally damaging alternatives to CFCs. It was Greenpeace that took the initiative in 1993 in developing greenfreeze, the first hydrocarbon-based refrigeration technology. Similarly, NGO involvement in campaigns against the dumping of Brent Spar (Dickson and McCulloch 1996) and against firms involved in genetic modification depended partly on strengthened links to scientists. The fierce arguments surrounding the objectivity of scientific advice reflected the wider problems of unidentifiable risk (Beck 1992), loss of trust in government and suspicions about the extent to which big companies were concerned about their environmental impacts.

Within government, science and technology has also come to play a stronger role. Licensing and regulatory agencies have turned increasingly to scientific knowledge when negotiating with firms over BATNEEC details or the carrying capacity of rivers. Similarly at the subnational level, policy-makers have relied on scientific expertise to help make decisions. In the early 1990s Kent County Council in south-east England established, as part of the EIA process, a computer programme that could analyse the impact of individual proposals for gas-fired power stations on air pollution levels. Similarly regulatory agencies use increasingly elaborate specialist knowledge when monitoring industrial operating processes.

Private sector influence on policy-making

Ecological modernisation has created new opportunities for private sector influence on policy-making. This reflects the fact that earlier approaches to environmental protection have not worked. Deteriorating environmental conditions are partly the responsibility of governments because of the implementation deficit, as discussed above. This analysis led state and industrial elites to conclude that the state's approach had to be reformed in ways that took greater account of private sector perspectives as to what was feasible. The promotion of ecological modernisation thus created new opportunities for big firms and trade associations to influence the development of policy, at the stage when the environment was becoming more central to government decision-making. The environment had thus become a more significant consideration within changed parameters. What emerged was a shift away from traditional bureaucratic top-down dirigism (Mol 1995: 362–3) to open-ended discussions which amounted to 'regulatory negotiation' (Weale 1992: 175).

However, it is important to distinguish the policy frameworks from the implementation details. Sometimes industry has a strong influence over the shape of the policy. On other occasions it is only really influential in terms of amending the implementation details. In practice this enhanced opportunity for influence has contributed to a variety of state–industry relationships along a strong state–weak state continuum. Power within the relationship can shift in two directions(- see Chapters 1, 3–7, 9 and 10). First, a weak neo-pluralist state allows regulatory negotiation to cover policy frameworks. An example here is industrial interests being able to veto EU carbon tax proposals and argue for voluntary agreements (Ikwue and Skea 1996). Such cases suggest that Lindblom's analysis (Lindblom 1977) of business interests being advantaged needs replacing with capture theory approaches. On the other hand, a stronger neo-corporatist state can retain control of the policy frameworks and restrict regulatory negotiations to details, thus imposing a policy strategy over time. Germany, Japan, the Netherlands, Norway and Sweden – the clean and green five as Dryzek refers to them – are examples here (1997: chap. 8; see also Weale 1992).

The range of state–industry relationships is well illustrated by changes in corporate environmental reporting in the 1980s and 1990s. Conventional financial accounts say nothing about a firm's impact on the environment. Drawing from earlier experiences in the 1970s (Estes 1976), companies began to develop accounting approaches focused on attempting to control and reduce their environmental impacts (Owen 1992; Gray 1994). Energy accounting was a common initial target. Rhone Poulenc, the chemicals giant, developed accounting systems to track its wastes and effluent discharges. BSO/Origin, the Dutch company, was one of the first to produce environmental accounts. Later, surveys of 500 big companies (*Financial Times* 14 September 1995), and later of 400 (Elkington 1996) showed how firms had moved on from just setting targets. They tried to quantify their environmental achievements. These reports were criticised by environmentalists and stakeholders for their glossy greenwash, their superficial analysis and their lack of transparency (Elkington 1996; *Financial Times* 17 April 1996). The Body Shop tried to develop more ambitious approaches to environmental auditing, analysing not just the company's ecological impacts on the biosphere, but also environmental conditions around major plants and the social impact of the company's activities on the workforce and other stakeholders. However, these initiatives remained voluntary. The great majority of firms of all sizes simply ignored them. The weak neo-pluralist state offered rhetoric and encouragement, but was anxious to avoid acting in ways that discouraged investment, especially by transnational corporations. By the mid-1990s though, a much stronger approach began to emerge in Denmark, Sweden and the Netherlands. Their governments began to introduce mandatory environmental reporting requirements. They had concluded that the voluntary approach would not work and that it was only by changing the framework, raising standards and forcing the pace, that the short-term profit maximisers and the rest would follow. This approach fitted more closely into the neo-corporatist model. Companies and trade associations were drawn into discussing the details, but not the principle.

A different kind of economic growth

The final feature concerns the nature of economic growth and builds from all the others. This is the way in which ecological modernisation aims to maintain economic growth while establishing 'a more environmentally benign development path' (Gouldson and Murphy 1997a: 75; see also Chapters 4–6, 9–10). It lays greater emphasis than in the past on protecting the environment and sustaining environmental conditions (Cairncross 1995). A more environmentally benign approach to economic growth involves moving away from a simple throughput model of the economy, with inputs of resources, materials and energy into the production process, leaving, as outputs, not just the product, but wastes of all kinds and lost energy. This does not reach Daly's model of a 'steady-state' economy (Daly 1992). But, in overall terms, ecological modernisation is different from conventional post-war economic growth, because it focuses on reducing resource consumption and generating less waste, 'while creating employment and improving economic welfare' (Gouldson and Murphy 1997a: 75; see also Chapters 9 and 10). At the site level this approach is referred to as industrial ecology (Marstrander 1994; Clift and Longley 1995). Kalundborg is one of the most frequently cited cases. It demonstrates how different processes can be integrated to create an industrial ecosystem on one site (Tibbs 1993). The integrating principle is that different producers of goods and services feed off each others' wastes. The following are all interlinked: an electric power plant, an oil refinery, a biotechnology plant, a plasterboard factory, a sulphuric acid plant, cement manufacture, horticulture, and district heating. As a result wastes and loss of energy are reduced to an absolute minimum; and interplant transport is scarcely needed.

This raises the issue of how ecological modernisation relates to the concept of sustainable development. Welford (1995); Elkington (1996), for example, broaden discussion about the scope of ecological modernisation, when arguing that it involves not just a corporate focus on profits and environmental quality, but on equity and an environmental ethic respectively, and on social justice too. The relationship between sustainable development and ecological modernisation is discussed further below. The point to establish here is that ecological modernisation is a more restricted attempt to reconceptualise the link between the environment and the economy. However this is not just about sustaining environmental conditions for further economic development. It also involves greater emphasis on environmental protection and environmental quality. This has important consequences at the local level in the neighbourhoods around major plants.

This more environmentally benign approach to economic growth emphasises the preventative, continuous reduction of environmental impacts strategy. This leads companies to improve sites, to clear up past environmental damage and to develop partnerships with local groups and schools. These and similar programmes contribute to enhanced quality of life for local stakeholders. This helps generate demands from local voters for higher quality environmental conditions. Subnational governments respond in developing programmes to protect and enhance the local environments in all sorts of ways (Roberts 1995; Selman 1996; see also Chapter 8). This is reinforced by people moving to such areas to benefit from the higher quality of life there. Weale uses the example of skilled workers in the new technology industries moving to places like Baden-Wurttemberg and Bavaria because of their environmental appeal (1992: 77). There are similar examples in Britain of people moving to the Welsh borders west of Birmingham and to the Pennine area between the Manchester and Leeds conurbations. Part of a different kind of growth is thus economic development that is compatible with enhanced quality of life. This is similarly reflected by the emphasis on clean environmental conditions in Denmark, and on southern California developing the strictest air quality regulations in the world (*International Herald Tribune* 13 September 1996).

The emergence of ecological modernisation

The origins of ecological modernisation can be dated to the late 1970s and the early 1980s. It was at this time that the new discourse was being discussed in three different supranational arenas, amongst three sets of elites. Hajer (1995: 96) shows how it developed along 'three different tracks'. The first was in a biodiversity setting where UN agencies and moderate NGOs like the International Union for the Conservation of Nature (IUCN) and WWF were drawn together. The *World*

18   S.C. Young

Conservation Strategy (IUCN 1980) focused on ways of conserving threatened habitats and ecosystems through 'efficient resource utilisation and considerate environmental planning' (Hajer 1995: 97; see also Chapter 3). Hajer's second track emerged from the debates in UN circles about the links between environment, development and security from the 1972 Stockholm conference on, through the UN commissions of the late 1970s and early 1980s, to the Brundtland Report of 1987 (Hajer 1995: 99–100). Brundtland aimed to shift perceptions of environmental topics away from their being seen as peripheral conservation issues, towards their being understood as a central dimension of core economic concerns (Baker *et al.* 1997). In the spheres of conservation, development and North–South relations, attempts were thus being made – largely at the supranational level – to establish links between the environment and economic issues.

The third track developed into a policy community around the OECD's Environment Committee. In the 1979–84 period, this operated largely as a thinktank, publishing reports, bringing together ministers and civil servants from the environment ministries that were being established in different countries, organising conferences and testing out the ideas emerging from academic networks (Hajer 1995: 97–9). This period culminated in the 1984 International Conference on Environment and Economics. This concluded that 'the environment and the economy, if properly managed, are mutually reinforcing; and are supportive of and supported by technological innovation' (OECD 1985: 10). Gouldson and Murphy (1997a: 74) later summed this up as follows: 'economic development and environmental protection can be combined to synergistic effect'.

In the mid-1980s the EU became a fourth supranational arena within which the core ideas of ecological modernisation were discussed. Officials from European countries that were the more advanced in environmental terms – the Netherlands and West Germany for example – were able to develop the emerging ideas further using the EU institutions. Weale (1992: 76–8; see also Chapter 3) points to the way in which the ideas were picked up and enlarged in the context of the EC's Fourth Environmental Action Programme, published in 1986. Mol shows in detail how EU policies on plastic wastes shifted from having virtually no environmental content in 1985, to a significant Council of Ministers resolution in 1987 (1995: 268–9). This resolution led to four environmental directives. Prior to its introducing the draft packaging directive, DG XI took the initiative in the late 1980s and set out to persuade polymer industry chiefs of the need either to recycle plastic waste, or to prevent it from being a problem via other means. The Association of Plastic Manufacturers in Europe welcomed this as a move to harmonise the different rules and targets being set in different countries. But discussing *where* ecological modernisation emerged, begs the question as to *why* it was attractive to different actors for different reasons.

The underlying reasons

The main reason usually given for governments being attracted to ecological modernisation is the implementation deficit problem discussed above. However, other contributory factors are involved too. In a number of respects ecological modernisation has improved the relationships between government and industry, making it less confrontational and more cooperative. Economic instruments hold out the promise of lower public spending on regulation and a slimmed-down, smaller state. This all fitted in well with the neo-liberal ideas that were promoted by think tanks in the 1980s and widely adopted by governments (Hajer 1995: 33). The pressures on member states continued into the 1990s (Chapters 1, 3, 4 and 7). By the 1990s, some 'realo' greens and many politicians on the left – initially suspicious of financial instruments – had come to accept the arguments about their flexibility and pragmatic value (Eckersley 1995; see also Chapters 1, 4 and 10). There are opportunities for governments to manipulate the new institutions being created and the established ones being adapted. In addition, corporate financial resources and professional expertise are being drawn into the emerging partnerships between companies and stakeholders, especially at the local level where there are benefits for subnational governments. When Nokia, the Finnish telecommunications firm, acquired a manufacturing plant at Bockum in Germany, for example, public transport access to the works was limited and commuting distances for employees averaged 50 km per employee. Nokia contributed about DM140,000 of the DM700,000 raised by the company, two local councils and the public transport operator to improve the rail infrastructure and service frequency (Wood 1995: 18).

In addition, ecological modernisation brings national and subnational governments benefits in the form of environmental gains – less polluted water courses, and the reuse and landscaping of damaged sites in ways that improve local biodiversity and the quality of life for local people (Selman 1996). There is also the prospect of further environmental benefits emerging. More widespread acceptance amongst producers of responsibility for post-consumer waste should help reduce the amounts of waste going to landfill and lower demand for new landfill sites.

Ecological modernisation was also attractive to governments and industry because it offered a way of appeasing and accommodating parts of the environmental movement, not to mention undermining the appeal of green parties. The earlier discussion showed how many NGOs had been moving away from their earlier confrontational strategy towards the more pragmatic approach of trying to influence policy. They began to argue their case in 'the appealing terms of ecological modernisation' (Hajer 1995: 102); and thus got drawn into the emerging discourse. From an NGO perspective, there were two main reasons lying behind this change of strategy in the 1980s (Weston 1989; Dalton and Keuchler 1990; McCormick 1991; see also Chapters 2 and 7). First, many of the NGO leaders and thinkers began to argue that they could achieve significant goals by getting involved in the debates. This created a situation in which they could push their own ideas; and try to amend government policies as they developed. Second, NGO leaders began to realise that they could use the issues of the late 1980s, like road building and ozone depletion, to build up their membership numbers. They argued that being able to demonstrate increased support would give them greater potential for influence, because they would be seen in political and industrial circles to be speaking for significant numbers of people. This shift from radical confrontation politics to discussion and compromise fitted in with the broader strategies being pursued by governments as part of their attempts to change the terms of the environmental debates and develop new solutions.

More broadly, ecological modernisation was attractive from the perspective of governments, because it addressed environmental issues without introducing the need for fundamental structural change. Drawing NGOs into the debates side-stepped the radical calls from the environmental movement and counterculture groups in the late 1960s and early 1970s for a complete restructuring of society and its institutions (Dalton and Keuchler 1990). In addition, ecological moder-nisation did not threaten late twentieth-century capitalism: it sought to adapt it (Weale 1992; Hajer 1995; Mol 1995; Blowers 1997). The pragmatic character of ecological modernisation is important, because it made it possible to build consensus and to avoid both damaging political disputes and the alienation of capital. This helped give the discourse wide appeal across different industrial sectors and draw in actors from civil society.

Moving on to industry's perspective, growing numbers of opinion leaders came to understand the core argument: *that more protection for the environment does not mean, as had been believed in the 1970s, less economic growth.* On the contrary, the approach leads to increased productivity gain from resources (Porter and van der Linde 1995). Economic growth that damages the environment and the suste-nance base undermines the prospects for future growth. Rivers and the air in cities can only continue to disperse wastes if amounts of pollutants are kept below specific levels. Industrial leaders had thus come to understand that it was in industry's interest to limit and control its impact on the environment – in order to safeguard its future prospects. It is important to stress that companies do not develop a broader, long-term interpretation of their environmental responsibil-ities for altruistic reasons. Their concern is economic self-interest. Operating within tighter environmental limits becomes commercial common sense.

This line of reasoning is reinforced by the commercial benefits that arise. Environmental concerns have produced demands for new products: more sophisticated pollution measuring equipment, environmental services and products like the non-polluting electric car. Even wastes become saleable assets. The declining interest in nuclear power led not just to interest in renewables, but to power stations being fuelled from such sources as rubber tyres and poultry litter (*Environmental Digest* 1996, 8: 5). The funding of such operations is a growth area for banks. Environmental concerns also create new marketing opportunities. One Scottish whisky was marketed in the early 1990s as the ultimate green product – combining clean, clear water with organic barley and pure yeast! Developing a green halo becomes a means of projecting an image as an environmentally responsible company. Such approaches help provide marketing opportunities, buttress stock market valuations, replace environmentally damaging products – thus regaining lost business; improving customer relations; and attracting new customers. They illustrate the arguments of writers like Schnaiberg (1997) that companies respond for economic reasons, although the argument here is that the reasons are more complex.

On some issues, environmental considerations can lead to increased costs. But these are not always a deterrent (see Chapters 1, 3–5, 7 and 10). Energy-saving measures, for example, require capital investment. But in the longer term they can automate away labour costs and generate revenue savings and thus reduce unit costs. Sometimes the initiative is taken within an industry and is, in effect, self-imposed. This occurs for example where the European Commission agrees draft directives with trade associations – as in the plastics case (Mol 1995). On other occasions, the costs of taking action to protect the environment in some way can be shared with other companies, with suppliers, retailers and other sectors. However, if one firm takes the lead, it can incur extra costs and be disadvantaged. But if all members of a trade association or the dominant firms in a sector act at roughly the same time in a competitive situation, then loss of competitiveness does not arise. The plastics and OMMRI cases show that the gains that flow from acting in an environmentally responsible way are thus often there for the taking.

Companies adopting the ecological modernisation approach can also gain pragmatic benefits. For example, improved relations with regulatory agencies and subnational governments can ease the burdens on managers. Better relations with stakeholders can help reduce public pressure for further regulation. New institutions provide opportunities for industry to influence emerging debates and the detail of new proposals. In addition, involvement in partnerships with other stakeholders at the local level can bring in extra resources. These range from external funding from the EU or national government to professional expertise and access to information exchange – as in the cases of recycling networks or landscape renewal.

The pressures that drove industrialists to adapt their approach also affected workers. During the debates of the late 1960s and 1970s, unions had put jobs before environmental concerns. In the 1980s and early 1990s their attitudes were affected by wider processes of change – spreading automation, deregulation of the labour market and post-Fordist change (Amin 1994). As ecological modernisation began to spread, it became apparent that the arguments that were being deployed in the boardroom were being used to persuade workers to go along with the need for change. Although it only happened in Belgium in the late 1990s (Chapter 1), in general, unions swung round behind the push towards ecological modernisation. The attitudes of workers were also affected by the broader processes of society's responses to the environmental change.

Moving on, ecological modernisation has also emerged because of growing public concern about the environment. This is sometimes understood simply in terms of the salience of the environment as an election issue, the growth of green parties, mainstream parties relating to the environment and new demands affecting institutions and programmes. The argument goes that issues like acid rain, the thinning of the ozone layer, the Chernobyl disaster, Exxon Valdez and urban smog received high profile media treatment during the mid and late 1980s and early 1990s. This coincided with the period between the publication of the Brundtland Report (WCED 1987) and the 1992 Rio Earth Summit (see Chapters 2–4 and 8). This approach gets linked to polling evidence to show how voters see the environment relative to jobs, crime, health, education and so on (Worcester 1997), to election results (Richardson and Rootes 1995) and to the environment being a mid-term issue (Carter 1992).

However, the increased salience of the environment as an issue after the mid-1980s needs to be more fully understood in the context of post-materialist value change. Ingelhart (1977) famously argued that the experience of those born after the Second World War was in sharp contrast to that of their parents and grandparents. They had been much more concerned with poverty, housing and trying to improve their material conditions to make life less of a struggle. Those growing up in the 1950s and 1960s, however – especially the better educated – attached greater importance to environmental and post-material issues during the 1970s, 1980s and 1990s. The way that these values affected peoples' behaviour is linked to the earlier discussion about stakeholders. Apart from their work, people are also individuals making decisions about where to live, and where to raise families; and as members of local communities around big plants, as members of local groups and non-green parties, as tourists wanting clean bathing water, as consumers demanding goods that do less damage to the environment, and as people concerned about the prospects for their families, friends and grandchildren.

The important point here is not how people voted, or how the environment went up and down political agendas. It is how changed values affected peoples' approaches on a range of topics beyond voting; and about how awareness of, and concern about, environmental issues reached higher levels than before. In the 1990s this became a higher base compared to the early 1980s (Worcester 1997). From this base, concern could erupt, as with the spontaneous boycott of Shell petrol stations in Germany after the Brent Spar incident; and the increased green vote in Belgium in 1999 after the food scares (see Chapter 1). The fact that the environment then fell down agendas, relative to crime or education or other issues, did not reduce that higher basic level of concern. It was this increased basic level of concern on environmental issues that was significant as part of the context from which ecological modernisation emerged. There was not a direct causal link between public attitudes and the emergence of ecological modernisation. But there was a symbiotic relationship between the two. Post-material values, as expressed through the higher levels of basic concern, helped sustain ecological modernisation, once it had became established, and even to accelerate its emergence.

Globalisation has also been a factor helping to promote ecological modernisation. Briefly this is the move away from relatively closed national economies with international links as a relatively minor feature. The situation has been transformed by the increasing value and volume of international transactions. National economies have become increasingly interdependent and a new global economy has emerged, reflecting wider post-Fordist changes (Amin 1994). This new global economy is characterised by increasingly integrated production networks which are worldwide in their scope and organised and controlled by transnational corporations (TNCs) (Amin and Thrift 1992). Other features include more extended trading patterns; open capital markets, as the finance and service sectors become more globalised; the establishment of a new hierarchy of world cities to serve the needs of TNCs (Friedmann and Wolff 1982; Knox and Taylor 1995); and increased cross-border capital flows, as growing numbers of TNCs relocate office functions to where specialist service are available, and plants to where labour is cheaper, and the labour market more flexible.

Economic aspects are linked to other factors which become part of the whole globalisation process. These include the growth and spread of electronic communications that leap across international boundaries (Sassen 1994); a concentration and restructuring of the media on a global basis; the emergence of elements of a common culture as the media, arts, films, sport and advertising all become increasingly globalised (Featherstone 1990); and the setting up of more supranational and intergovernmental organisations involving the establishment of global policy-making regimes, with, it is widely argued, a reduced role for the traditional nation state. The local impacts of globalisation thus reflect the interaction of economic trends, political events, cultural processes and other factors that have their origins in different parts of the planet.

Ecological modernisation is closely linked to globalisation (Chapter 9). Without getting into the debates about its nature and extent (Giddens 1990; Robertson 1992; Hirst and Thompson 1996), the point here is that it intensifies competition. The liberalisation of markets and the reduction of trading barriers promote the ubiquitous level playing field, thus 'rendering the notion of comparative advantage obsolete' (Porter and van der Linde 1995: 132). Sectors like telecommunication and energy have been transformed by competition linked to innovation and technical change. In addition, the countries and the TNCs that establish the highest environmental standards force others to follow them (Weale 1992). In the car industry, for example, in the early 1990s, Japanese firms built cars to higher environmental specifications than required by the countries to which they exported them. Other TNCs and companies then followed. They believed that in responding they were reflecting changes in consumer demand for increasingly higher environmental standards and moving into an expanding market. Companies that resisted innovation lost competitive advantage (Porter and van der Linde 1995). Similarly, Mol highlights the way ecological modernisation and globalisation have reinforced each other in chemicals (1995: 375–7). Globalisation is also a factor in promoting ecological modernisation where there are links between sectors, such as in the cases of the firms that banks fund and those supplying the food retailers. In a variety of ways globalisation thus helps to establish and then spread ecological modernisation.

The emergence of ecological modernisation

The concept emerged initially in four separate policy communities around the IUCN, the UN, OECD and the EU. They were linked to national governments and industrial leaders. Ecological modernisation was initially pushed forward by policy entrepreneurs within and between these arenas and fuelled by globalisation. It also maintained momentum partly because of heightened environmental concerns and post-materialist values. NGOs played a minor, reinforcing role. Hajer (Hajer 1995: 100) shows how the development of 'a common vocabulary' helped to draw together 'a coalition of forces'. He rightly emphasises the way ecological modernisation emerged as the dominant policy discourse, rapidly conquering the environmental domain (see also Chapter 9). It proved a coherent enough concept to draw a positive response from senior decision-makers in industry and from state elites (see Chapters 1, 3, 4 and 10).

The willing involvement of industry in particular was central. A critical aspect of this was that – as discussed above – ecological modernisation did not threaten contemporary capitalism. Ecological modernisation can best be understood as a late twentieth-century strategy to adapt capitalism to the environmental challenge, thus strengthening it. The contrast with the attempts at indicative economic and industrial planning in Western Europe in the 1960s and early 1970s, backed by proactive state intervention, is illuminating here (Young 1974; Hayward and Watson 1975). These planning initiatives had an erratic impact. Although they were initially welcomed by some industrialists, most responses in the private sector varied from passive acceptance to outright opposition. In practice the overwhelming majority of firms were able to operate quite independently of government and public sector planning initiatives. With ecological modernisation though, firms responded positively to the ideas while retaining their autonomy. In particular the leading role of TNCs was important here. They had been a problem in the earlier planning era, often being attacked for undermining state initiatives. In the era of ecological modernisation, they were at the forefront of the push for a new approach. In addition the way the discourse of ecological modernisation drew strongly from the language of industry, the discussions in the OECD arena in particular led to the idea being presented to industry in terms of pollution representing inefficiency, and conventional approaches wasting resources that could be exploited more profitably (Hajer 1995: 31).

The moves in industry towards supplementing conventional financial accounts with environmental reports provide an interesting example of how these different factors interacted with each other. These reports cover such things as environmental audits, environmental contingent liabilities and the extent to which the company has met the environmental performance standards imposed by governments. Gray (1994) shows how environmental disclosure is increasingly being taken up in North America and western Europe by TNCs and nationally-based companies, as a result of initiatives from the UN Centre for TNCs and the EU combining with response from finance directors, state-industry advisory bodies and the accountancy profession in ways that reflect more pragmatic NGO attitudes and changing social values. Environmental reporting began on a voluntary basis. But Gray argues that regulatory initiatives like the American Superfund Act and the EU's proposals for a European Superfund Act are establishing it on a more widespread and permanent basis. LCA provides another example of ecological modernisation being promoted through the interaction of the different factors. Berkout (1996) shows the influence of government regulations designed to change end-of-life wastes, and industry reacting to innovation by competitors and to changing consumer demands. This is all in the wider context of a UN Environmental Initiative (Clift and Longley 1995: chap. 6). Here again, ecological modernisation is being promoted not by specific causes like an OECD report, government regulation or changed consumer values, but by the interaction of the different sets of factors. Cases like these reveal a strong coalition of support for the concept.

Ecological modernisation in the 1990s

During the 1990s, the discourse of ecological modernisation became more widespread, and its main features emerged. However, although there are lots of examples of the eight features highlighted above, the extent to which each of them is effectively and widely established is unclear. For example, economic instruments were increasingly applied in Denmark but, despite much debate from 1990 onwards, little used in Britain (see Chapters 4 and 7). By early 1996, concerns were expressed at an OECD review meeting about the lack of progress on integrating the environmental and economic dimensions, and on extending economic instruments in agriculture, energy and transport (*Environmental Digest* 1996, 2: 2 and 7).

In geographical terms it seems to have made most progress in Western Europe, as in the Netherlands, Germany and Scandinavia (Weale 1992; Hajer 1995; Gouldson and Murphy 1997a; see also Chapters 1, 3–5). But, as the examples quoted show, some features are present in the USA, Canada and Australia. Hillary (1997) adds case studies from New Zealand, Brazil, South Africa, Hong Kong, Eastern and Central Europe and even China. Christoff argues that it is mainly a European phenomenon (Chapter 9). However, many of these cases are centred around isolated corporate-led examples. There are many examples of some features existing within some countries and in some regions. Often though, the features are not collectively present, and are not clearly established in the context of collaborative policy-making partnerhsips involving government, industry, moderate environmentalists and scientists – as in the case of Dryzek's 'clean and green five' discussed above.

It is difficult, looking across the OECD countries and bearing the global reach of multinationals in mind to say how far the example of the pioneering companies have been followed. There has certainly been an impact in such sectors as paper, vehicles and chemicals. These examples are usually linked to big companies and TNCs, but increasingly these and other authors, like those above, are pointing to small and medium-sized enterprises (SMEs) which have grown because they have put ecological modernisation at the heart of their investment strategies. It is also apparent that many companies of all sizes still operate on the basis of short-term, profit maximisation, irrespective of environmental damage. Multinationals are able to relocate to other countries if the regulatory framework makes too big an impact on their business strategy, thus simply shifting environmental degradation around (Gouldson and Murphy 1997a; see also Chapter 9). The talk of win–win situations and the highlighting of good news stories on the corporate environmental front remains focused on a fairly small group of pioneers and followers. But cases like the oil companies diversifying into renewable energy give credibility to the idea of big firms moving on from a few environment-friendly products to changing their core business activities, operating more within the biosphere's limits and maintaining profitability. Gouldson and Murphy (1997b: 77) review a series of studies before concluding: 'Any ecological modernisation that has occurred so far has not been on a sufficient scale to back up the claims of its advocates that it offers a genuine escape from the environmental problem'. The question for environmentalists is whether ecological modernisation is an effective strategy for promoting sustainable development and tackling the whole range of problems that face the biosphere (Welford 1995).

The relationship between ecological modernisation and sustainable development

Many writers have pointed to the ambiguity and contestability of sustainable development (Redclift 1993; Beckerman 1994; Kirby 1995; Reid 1995; Welford 1995). Analysing the different dimensions of ecological modernisation in terms of different understandings of sustainable development helps reveal the nature of ecological modernisation. Baker *et al.* (1997: 8–18) have used the idea of a ladder as a heuristic device to convey the range of interpretations of sustainable development and the policy implications associated with it. Four rungs are analysed in terms of nine features. These include the roles of the economy, nature and technology, and types of institutions.

At the foot of the ladder is the treadmill approach to post-war capitalism. Little attention is paid to environmental concerns. This reflects the earlier discussion about environmental controls being seen in the 1970s as extra costs, so that choice became jobs or the environment. On the rung of weak sustainable development more attention is paid to environmental concerns (Pearce *et al.* 1989). But conventional economic growth remains the aim. In contrast though, strong sustainable development sees environmental protection as a precondition for the different kind of economic development. This third rung is much more focused on the environmental dimension. At the top is the, largely unobtainable, ideal model of sustainable development. This is based on the views of writers like Daly (1992). It takes an ecocentric world view, stresses holistic approaches that include the social dimensions of sustainable development and seeks profound institutional and structural changes in society.

Ecological modernisation has a number of features in common with the treadmill model. It is strongly anthropocentric and has a managerialist approach to the environment and to dominating nature. It is strongly technocentric, using technology to get around finite limits to growth (Hajer 1995: 32). It accepts globalisation and the benefits of the market system, and it sees no need to curb consumption. The mobility of capital investment is wasteful in terms of how it affects labour and resources (Hajer 1995: 32).

Ecological modernisation also shares certain features with weak sustainable development. It accepts the importance of futurity and of future generations having access to similar resources to existing generations. It accepts the need to consider environmental costs in decision-making and not just the economic costs. It accepts the need in general to control pollution. It also accepts the need where possible to make development compatible with local ecosystems. But it believes that where preserving critical natural capital conflicts with high economic priorities, the latter should come first.

In some respects though, ecological modernisation does get beyond the tokenism of weak sustainable development. In seeking to reconceptualise the links between economic growth and the environment, it has developed some features in common with strong sustainable development. It stresses clean technology, innovation and the need to get beyond end-of-pipe approaches. It stresses environment-friendly products and supply chains. It looks for opportunities to integrate different sectors, as with energy from waste. It accepts the needs to restructure institutions and to develop more sophisticated sets of market-based instruments. It questions exponential economic growth and seeks a qualitatively different development model that also stresses the need for environmental quality for local stakeholders. The environment is no longer just a sink for wastes, a free good. It is valued in its own right, because it provides the sustenance base and enhances the quality of life.

Ecological modernisation does not fit neatly with one approach to sustainable development. Further it actually rejects some features of sustainable development. Its links to globalisation and the World Trade Organisation's attempts to reduce trade barriers (*Economist* 1 June 1997: 108) mean that it does not relate to serving needs rather than wants, nor to intragenerational equity, social justice or distributional issues. It is ambivalent on participation, taking a fairly technocratic, top-down approach to community involvement (Hajer 1995: 78–89), while accepting the need to give a stronger role to NGOs in order to reduce conflicts (Weale 1992: chap. 6). It does not relate to local economic self-sufficiency or to the need to develop modern forms of bioregionalism (Welford 1995; Dobson 1995).

The heart of ecological modernisation

Linking the features of ecological modernisation to different perspectives on sustainable development highlights the central dynamic of ecological modernisation. It is late twentieth-century capitalism's response to the emerging environmental challenge. Dryzek sees it as 'the ecological restructuring of capitalism' (1997: 145). For Gouldson and Murphy, it is 'a means by which capitalism can accommodate the environmental challenge' (1997a: 75). Governments have acquiesced, agreeing to go along with it and actually promoting it, because, as argued above, it addresses the environment problem while avoiding the need for structural economic change. Industry has responded extensively on some issues and hardly at all on others. This explains why ecological modernisation has some features in common with strong sustainable development, some with weak and some with treadmill capitalism. As a system, capitalism has responded as far as it needs to in order to survive and continue to prosper in the short term. It has exploited new opportunities and new markets. It has responded to new regulatory pressures. It has moved to accommodate increased costs on things that range from detailed environmental assessments of new projects and greater planning inquiry costs, through to pollution control equipment. In addition, growing numbers of firms of all sizes have come to appreciate the increasing importance of projecting and polishing their green halos. This is because reputation and image affect sales, and ultimately stock market valuations. Late twentieth-century capitalism has thus come to terms with the need to pay greater attention to waste. This can range from the disposal of oil rigs like Brent Spar to addressing the company's responsibility for post-consumer waste – as with engine oil. Schnaiberg (1997: 76) picks up this dimension of public relations and image, arguing that capitalism's response 'appears to be more a social construction put on disparate corporate activities and motives'.

The ambiguity of ecological modernisation centres ultimately around writers' different perspectives, if not values. Schnaiberg would argue that it is a minimalist response by capital to the environmental critique. On the other hand, writers like Welford would suggest that it is starting to reflect a new, more responsible, environmental ethic. Industry perceives the framework within which it operates as threatened. Ultimately its future prosperity and growth is at stake. But whichever view is taken, there is, at the heart of ecological modernisation, a dawning realisation that continuing economic success can only be achieved over the longer term if industry responds to the environmental agenda, accepts widened responsibilities for environmental problems and adapts its behaviour accordingly.

The future of ecological modernisation

The prospects for ecological modernisation are confused. The factors explaining its emergence remain in place as a set of pressures promoting its growth around the turn of the century and beyond. However, its wider establishment is not inevitable. Despite its spread, there are strong forces working against its development. The contributors to this book provide detailed examples of how conflicts undermine moves towards ecological modernisation: conflicts between different groups and governmental actors (Chapter 1), conflicts between government departments (Chapters 2 and 7), conflicts between environmental programmes and the promotion of competition policies and economic growth (Chapters 3 and 6) and conflicts within governments over the extent to which economic instruments should be used and subsidies to smokestack industries phased out (Chapters 4 and 5).

While globalisation can promote ecological modernisation, the mobility of capital and the short-term interests of TNCs can also work against it. Multinationals relocate not just to achieve savings on labour costs, as in south-east Asia and access to new markets, as in eastern Europe. They are still able, apparently, to find countries where environmental standards are lower (Barnet and Cavanagh 1994). A number of writers argue that, although there are exceptions, ecological initiatives and ethical arguments are almost always crowded out by market pressures and other commercial considerations (Gould *et al.* 1995; Walley and Whitehead 1994). Mol (1995: 360) also highlights cases where economic criteria have overruled ecological criteria, as in chemicals, which has been one of the leading sectors. The short-term vested interests of big companies sometimes express themselves through the 'backlash' of strong industry groups arguing that the environmental critique is flawed and exaggerated (Rowell 1996). A good example is the Global Climate Coalition's determined lobbying against the Framework Convention on Climate Change negotiations in the mid-1990s ahead of Kyoto in 1997. It represented TNCs and American fossil-dependent companies. The Australian Industry Greenhouse Network was a similar example (*International Herald Tribune* 12 July 1996).

Furthermore, some of the features of ecological modernisation turn out to be less than originally envisaged. Berkout (1996: 154) stresses that, in practice, LCA has a number of limitations: 'The assumption that LCA can be pervasively internalised as an environmental management tool is unfounded'. One of the main problems is that 'claims of life cycle environmental benefits are not safe because they inevitably lead to counter-claims, also LCA based' (Berkout 1996: 153) – see also Kluppel (1995). This occurred in the arguments about the environmental merits of washable nappies as against disposable ones. The problem arises because it is difficult to determine where the boundaries of what is being analysed should be fixed. Welford argues this has been especially difficult for agrochemical, petrochemical, chemical and mining companies: 'A full ecological consideration of product life-cycles [...] has to take into account the impact of raw materials procurement on biodiversity, endangered habitats, human and animal rights and non-renewable resources' (Welford 1996: 214). This leads Berkout to conclude that the early enthusiasm for LCA in the context of ecological modernisation is waning, especially in chemicals. Apart from not giving clear guidance, there are also fears, especially in America, that LCAs will open firms up to liability claims.

On the other hand though, there are factors favouring the continued spread of ecological modernisation in the industrialised democracies. These are related to the theoretical arguments in favour of ecological modernisation from the perspectives of both government and industry, as discussed above. Certainly, the discourse of ecological modernisation has proved it has wide appeal and can attract an initial coalition of support. But the further expansion of ecological modernisation will depend in part on how far industrial and commercial elites are convinced by evidence emerging that there are realisable advantages in it for them. Thus globalisation will promote ecological modernisation if it helps produce demonstrable savings, greater resource exploitation and higher profits. Similarly, pressures within sectors will have an increased impact if they can be seen to be producing benefits. Consultants, sectoral trade associations and EU directives are all likely to influence companies left resisting or wavering about engaging with ecological modernisation. Again at the level of the firm, decision-makers will be moved by evidence not theoretical advantages. Cases like the production of solar panels going up by 42 per cent in the US during 1996–7 in response to a surge in demand, will help persuade dubious industrialists that there really is something in the environment for them (*Greenpeace Business* February 1998: 5).

Political leaders at national and subnational levels have also been faced by largely theoretical arguments about the advantages of economic instruments and other features of ecological modernisation. But the evidence of quick impact has been limited (see Chapters 5, 6 and 7). However, when good news filters through, it does offer encouragement. It emerged in 1996 that the environmental protection sector in Germany employed nearly a million people – which was almost as many as its successful vehicle manufacturing industry (*Environmental Digest* 1996, 10: 7). Evidence of clear advantages like this will encourage governments to resolve the kinds of conflicts between departments and policy programmes that this book reveals to have been widespread in the early 1990s. The commitment of political leaders to ecological modernisation will also depend in part on how far they are pushed by NGOs and industry; and on how far the post-material values, as embodied in the higher basic levels of environmental awareness discussed, will raise expectations and make further demands.

The prospects for ecological modernisation are also linked to its relationship with sustainable development. There are different views as to whether ecological modernisation has a positive effect on the promotion of sustainable development. The radicals' view is that ecological modernisation undermines strategies to promote sustainable development. Critics warn that it is an attempt 'to legitimise and sustain the very structures and systems that have been responsible for environmental decline' (Gouldson and Murphy 1997a: 75). It assumes that technofixes will work. It does nothing to limit advertising or reduce consumption levels (Dobson 1995) or move us away from globalisation towards the development of modern forms of bioregionalism. It has nothing to offer Third World countries. Indeed some environmental problems have been, and are likely to continue to be, exported to these countries(see Chapter 9). Moreover, as it is only being pursued by a minority of companies, predominantly in the industrialised democracies, it has little chance of making serious inroads on the scale of the biosphere's environmental problems. The environmental gains are cancelled out by economic growth (Chapter 5). Meanwhile the impact of the post-Rio attempts to influence the mass of companies beyond the pioneers appears limited. Amongst those who argue this position are those who see it as capitalism's holding strategy, an attempt to buy time. Hajer (1995: 33–4) refers to the view that ecological modernisation is a 'wolf in sheep's clothing'. He asks whether it is 'in fact a rhetorical ploy that tries to reconcile the irreconcilable (environment and development) only to take the wind out of the sails of the "real" environmentalists'.

The pragmatists, on the other hand, have a different view of the extent to which ecological modernisation helps promote sustainable development. For them it is the least bad option. From the perspective of strong sustainable development, there may be limitations to the promising initiatives outlined by writers like Cairncross (1995); Roberts (1995). Nevertheless, the earlier discussion shows how clean technologies lead to firms' improving their performance and to growth and expansion. The pragmatists argue that in promoting best practice approaches like green technology and eco-efficiency, writers like Schmidheiny (1992) are giving us a useful compass bearing. They are starting us in a positive direction on a route across the sea towards strong sustainable development. Ecological modernisation does have some features of strong sustainable development. But there is the problem that, as yet, the surveys have only been done for the early part of the journey.

Meanwhile, the pragmatists see the radicals as being unrealistic in expecting too much too soon. For the pragmatists the issue is best understood via an extension of the marine analogy. An oil tanker speeding along on a set course cannot turn round quickly. It has to be turned round in two phases. Phase 1 is about slowing down, reconsidering the route and getting pilots aboard. Phase 2 comes after adjustments have been made to the course being steered. But it is difficult to know what course to steer if the charts are not up to date, a new oil rig has been erected somewhere ahead and there is uncertainty over rising sea levels.

Pragmatists take a similar view of strategies to save the biosphere. It cannot be done quickly, as the radicals often imply. But it can be done in two phases or stages. Phase 1 would cover the process of getting growth-oriented economies under control. This is roughly equivalent to a weak sustainability position. This could take 10 or 20 years according to different views about controlling vested interests. We are, as Porter and van der Linde (1995: 127) put it, 'in a transitional phase of industrial history in which companies are still inexperienced in handling environmental issues creatively'. Phase 2 would then see the establishment of what Eckersley (1992: 321) calls the green market economy:

> a predominantly private sector market economy that is much more heavily circumscribed than is presently the case in most western economies. In particular, Green economists envisage that all market activity (whether carried out by the state or the private sector) should be more heavily regulated, scaled down in terms of material-energy throughput and made more responsive to ecological considerations and *informed* consumer preferences.
> (her emphasis; see also Chapter 10).

This seems roughly equivalent to a strong sustainable development position in the ladder model referred to above (Baker *et al.* 1997).

The difference between Phases 1 and 2 is one of scale. It is easy to envisage how in Phase 1 pioneering initiatives and good practice schemes – as with ecological modernisation – are becoming established. But initially they will be isolated and relatively insignificant in complex industrialised societies within a globalised economy. It is only when such approaches become widespread across whole economies and a critical mass emerges that Phase 2 will have arrived. In this view, ecological modernisation has a value, because it takes us well into Phase 1 while research and surveys are done to try to work out the route to Phase 2. If ecological modernisation has a value, then the next question to arise is how can it best be pushed?

Promoting ecological modernisation

There are two views as to whether ecological modernisation can best be promoted within OECD countries by industry or by the state. First, there are those who argue that industry should lead (Hawken 1994) and do so within a globalised neo-liberal economic framework (Cairncross 1991,1995; Schmidheiny 1992; Marstrander, 1994). They argue that then companies of all sizes will adopt ecological modernisation for economic and commercial reasons. Market forces will create a critical mass so that it will come to be in the interests of the resisters and short-term profit maximisers to change. Evidence here is provided by the chemicals case where small numbers of big companies have created pressures to raise standards, sometimes on a voluntary basis (Mol 1995; Aaron 1997). This approach implies that the neo-pluralist model of the state discussed earlier is the most appropriate.

However, other critics are sceptical of industry's claims (Welford 1996; Dryzek 1997: chap. 8; Gouldson and Murphy 1997a,b), arguing that industry's environmental performance can be improved more swiftly by state intervention and tighter regulation and that the state has no alternative but to lead if adequate safeguards are to be maintained. Such writers argue that a strategy of leaving it to industry will not work because the influence of the market will in practice, be more restricted; and that laggard firms can be made to act via a tightened regulatory framework (Porter and van der Linde 1995). While incentives can be used to tempt some companies, stronger measures will be needed to get the real resisters to change their approach. The other argument in favour of the state taking a more proactive role in leading, is that only the state can be relied on to continue systematically to winch the process of adapting ecological modernisation up to higher and higher environmental standards.

The factors working against change are strong collectively. The speed with which ecological modernisation becomes widespread will turn not just on the reaction of leading companies, but on the role of the state in cajoling and persuading the mass of companies that are not in the vanguard. During Phase 1, key factors will be the extent to which OECD governments are prepared to play a stronger role at both national and subnational levels; whether collective political will exists over time; and whether collaborative, consensus-based partnership approaches can generate change (see Chapters 1, 3, 4, 8 and 9).

In this context, the capacity of the corporatist model to promote collaborative approaches is illuminating. It is usually associated with its potential in the 1960s and 1970s to deliver income redistribution and welfare programmes (Katzenstein 1985; Williamson 1989). However, Dryzek (1997: 141) emphasises how the 'clean and green five' are all 'to greater or lesser degrees, corporatist systems'. In these countries, governments, businesses, moderate environmentalists and scientists have collaborated. Each country has a 'political-economic system where consensual relationships among key actors prevail'. He highlights in particular the way that the German case shows how influential environmental values become, once they have been adopted by key actors, within a collaborative policy-making framework (contrast with Chapters 1, 4, 6 and 9).

This approach to ecological modernisation requires a 'consensual and interventionist policy style consistent with corporatism' (Dryzek 1997: 151). Crucially, it is collaborative and non-adversarial. It rejects not just an industrial policy based on state intervention and public ownership in the pattern of the 1960s and 1970s in Europe, but also the deregulated market model that predominated in the 1980s. The state appears to play a more limited role than in the 1970s. But it establishes supportive policy frameworks, and promotes consensus-based solutions. Industry adopts ecologically responsible and innovative approaches. A neo-corporatist approach seems to offer a different strategy for promoting both economic prosperity and environmental protection.

However, this line of argument assumes that all parts of industry – and not just the enthusiasts and the pioneers – will respond. In practice though there is a long tail of laggards, resisters and short-term profit maximisers. It seems clear that, as Gouldson and Murphy (1997a: 81) put it, 'existing approaches to environmental policy fail to establish either a sufficient incentive or an adequate imperative to overcome the barriers that restrict change'. As a result the state has to be prepared, when collaborative approaches do not produce change, to lead. It needs to be able to stand outside the arguments and to determine environmental goals. It has to be prepared and capable of playing an informed, guiding and steering role. It has, on occasion, to impose its will, and drive regulation on in circumstances where leaving the promotion of higher environmental standards to industry is not producing results on a wide enough scale.

Playing more of a leading role implies the adoption of a forceful, interventionist form of neo-corporatism to promote state-led ecological modernisation in the OECD countries during Phase 1. It is based around collaborative approaches, and the increasingly imaginative use of incentives rather than inflexible regulations, to change corporate behaviour. The state aims to construct and maintain policy frameworks that progressively raise standards, while taking reasonable account of what is feasible from industry's perspective. But, different actors know that there is the implicit threat that the state is prepared to use its powers when necessary. This is also implied by Dryzek (1997: 141) arguing that 'conscious and co-ordinated intervention is needed to bring the required changes about'. The state can fall back on using its full powers proactively in two ways. First its powers can be exerted in cases where standards need driving up faster than industry thinks is necessary. Secondly, the earlier distinction between allowing industry to help reshape the policy framework, and restricting it to negotiations over implementation details within the policy frameworks, is relevant here. There are occasions where the state can insist on imposing the framework, while leaving the details open to discussion. This all has to be done in a self-critical, reflexive way (Mol 1995; see also Chapter 9).

If the state is to develop this forceful neo-corporatist approach and be more proactive in leading, it has to adapt its roles. Its approach to infrastructure provision needs to be reviewed in the context of the need for less travel. Issues like clean technology and reduced waste demand that specialist, 'hands-on' advisory services be extended and aimed more specifically at the tail of resisting and laggard companies (Gouldson and Murphy 1997a: 81–3). Next, the state has to play a sweeper role and be more proactive in tackling issues neglected by other actors – like clearing up contaminated land where the owner is bankrupt. National frameworks need to be developed in ways that can more readily be adapted to different subnational conditions (see Chapter 8). Next, the state needs to play more of a monitoring role, so that it can amend its programmes in the light of new developments and act swiftly to limit environmentally damaging activities. Finally, these roles need to be supplemented by a more extensive and ambitious approach to environmental education.

The most important change though is the need to develop economic instruments and regulatory controls in more flexible and sophisticated ways (see, in particular, Chapter 4). The skill and imagination with which the incentives are deployed and developed becomes critical. State-led ecological modernisation will only emerge on a broad front, reaching the mass of companies beyond the pioneers, if tempting incentives are seductively and persuasively presented to industry so that the resistance of the laggards and short-term profit maximisers is overcome. It will, for example, be necessary to remove environmentally damaging subsidies (as to oil and coal – see Chapter 5) and create incentives (as in the cases of the Japanese and American solar roof-top programmes and the Danish and Dutch off-shore wind energy equipment) across a much broader front if there is to be a real change of gear during Phase 1 in the OECD countries. Otherwise, the impact of this attempt to green capitalism on the future of the biosphere will be undermined, and limited to the contribution just of the pioneering companies.

Future research

This book shows the patchy spread of ecological modernisation. One of the main problems with research on ecological modernisation is that it is often approached via case-studies of pioneering initiatives. Several lines of research would help to bring the early studies into perspective so their significance can be more fully assessed. Some of the research topics listed here are sector based, some focus on geographical areas, and some shift the focus from theoretical and empirical analysis to the search for normative models. There is a lot of opportunity for comparative public policy approaches, not just in terms of policy-making and implementation problems, but also in terms of policy transfer.

1. To what extent and why does the promotion and practice of ecological modernisation vary from sector to sector? This introduction has highlighted how sectors like paper and chemicals become recurring examples. To what extent can policy-makers relate ecological modernisation to services, agriculture and non-manufacturing sectors? What is its potential significance in the context of whole economies?

2. How widespread is the broadening view of a company's environmental responsibilities? The first feature of ecological modernisation showed how this now stretches from the behaviour of its suppliers to the treatment of post-consumer waste. It involves the development of a new corporate culture that takes a more holistic view and has ethical dimensions.

3. To what extent can these new approaches be sustained over 10 or 20 years in the context of globalisation, technological change, firms seeking to reduce their net environmental impacts over time, and other factors? Monitoring studies need to be established. The selective intervention programmes of the 1970s appeared successful at first, but were subsequently largely reversed in a wave of neo-liberal criticism of their negative impacts.

4. From the point of view of the state, with its wider perspectives than industry, what are the most effective economic instruments and regulatory approaches? How do they vary in their impact from sector to sector? These issues need to be researched in the context of the earlier distinction between the state establishing the framework, and industry helping to shape and implement the details.

5. What is the scale and scope of the institutional changes and partnerships that are developing – from the subnational up through the national to the supranational level – between private sector and state organisations? How do these affect existing channels of influence between governments at different levels, trade associations and other industry lobbying groups, TNCs, small and medium-sized enterprises and local stakeholders? To what extent is the balance of power between these actors shifting away from industry?

6. At the subnational level, how is the concept of bioregionalism being developed to promote models of greater self-reliance or even greater local economic self-sufficiency? Both in peripheral regions, and in areas where smokestack industries have declined, subnational governments are promoting a variety of models to counter the way globalisation and other factors undermine the local economy. To what extent does ecological modernisation undermine local economies? Or can localised versions be developed that strengthen local economies – as with the green technology park strategy?

7. What is the extent to which ecological modernisation applies outside Western Europe? There are examples from Canada, the United States and Japan. But how far is the model established and relevant to the other OECD countries, the more developed post-Soviet economies like Hungary and the more successful of the Third World countries?

References

Amin, A. (1994) *Post-Fordism: a Reader*, Oxford: Blackwell.

Amin, A. and Thrift, N. (1992) 'Neo-Marshallian nodes in global networks', *International Journal of Urban and Regional Research* 16, 4: 571–587.

Aaron, S. (1997) 'The integrated approach: the Chemical Industries Association responsible care', in R. Hillary (ed.) *Environmental Management Systems and Cleaner Production*, Chichester: Wiley.

Andersen, M. S. (1994) *Governance by Green Taxes: Making Pollution Prevention Pay*, Manchester: Manchester University Press.

Baker, S., Kousis, M., Richardson, D. and Young, S. C. (eds) (1997) *The Politics of Sustainable Development*, London: Routledge.

Barnet, R. J. and Cavanagh, J. (1994) *Global Dreams: Imperial Corporations and the New World Order*, New York: Simon and Schuster.

Beck, U. (1992) *Risk Society: Towards a New Modernity*, London: Sage.

Beck, U., Giddens, A. and Lash, S. (1994) *Reflexive Modernization: Politics, Tradition and Aesthetics in the Modern Social Order*, Cambridge: Polity Press.

Berkout, F. (1996) 'Life cycle assessment and innovation in large firms', *Business Strategy and the Environment* 5, 3: 145–55.

Beaumont, J., Pedersen, L. and Whitaker, B. (1993) *Managing the Environment*, London: Butterworth-Heineman.

Beckerman, W. (1994) 'Sustainable development: is it a useful concept?', *Environmental Values* 3, 4: 191–209.

Blowers, A. (1997) 'Environmental policy: ecological modernisation or the risk society?', *Urban Studies* 34, 5–6: 845–71.

Cairncross, F. (1991) *Costing the Earth*, London: Economist Books.

—— (1995) *Green, Inc.: a Guide to Business and the Environment*, London: Earthscan.

Carter, N. (1992) 'Whatever happened to the environment? The British general election of 1992', *Environmental Politics* 3, 1: 442–8.

Clift, R. and Longley, A. (1995) *Clean Technology and the Environment*, London: Blackie.

Daly, H. (1992) *Steady State Economics, 2nd edn.*, London: Earthscan.

Dalton, R. J. and Keuchler, M. (eds) (1990) *Challenging the Political Order*, Cambridge: Polity Press.

Dickson, L. and McCulloch, A. (1996) 'Shell, the Brent Spar and Greenpeace: a doomed tryst', *Environmental Politics* 5, 1: 122–9.

Dobson, A. (1995) *Green Political Thought, 2nd edn.*, London: Routledge.

Dryzek, J.S. (1997) *The Politics of the Earth: Environmental Discourses*, Oxford: Oxford University Press.

Eckersley, R. (1992) 'Green versus ecosocialist economic programmes: the market rules OK?', *Political Studies* 40, 2: 315–33.

Eckersley, R. (ed.) (1995) *Markets, the State and the Environment*, Basingstoke: Macmillan.

Elkington, J. (1987), *The Green Capitalists: Industry's Search for Environmental Excellence*, London: Gollancz.

– (1996) *Engaging Stakeholders*, London: Sustainability.

ETA (Environmental Transport Association) (1995) *Car Buyers' Guide*, Weybridge: ETA.

Estes, R. W. (1976) *Corporate Social Accounting*, London: Wiley.

Featherstone, M. (ed.) (1990) *Global Culture: Nationalism, Globalisation and Modernity*, Newbury Park, CA: Sage.

Fischer, K. and Schot, J. (1993) *Environmental Strategies for Industry*, USA: Island Press.

Freeman, P. K. and Kunreuther, H. (1997) *Managing Environmental Risk Through Insurance*, Dordrecht: Kluwer.

Friedmann, J. and Wolff, G. (1982) 'World city formation: an agenda for research and action', *International Journal of Urban and Regional Research* 6, 3: 309–44.

Giddens, A. (1990) *The Consequences of Modernity*, Cambridge: Polity Press.

Gould, K. A., Schnaiberg, A. and Weinberg, A. S. (1995) 'Natural resource use in a transnational treadmill', *Humbolt Journal of Social Relations* 21, 1: 61–93.

Gouldson, A. and Murphy, J. (1997a) 'Ecological modernisation: restructuring industrial economies', in M. Jacobs (ed.), *Greening the Millennium: the New Politics of the Environment*, Oxford: Blackwell.

— (1997b) *Environmental Policy as Practice*, London: Earthscan.

Gray, R. (1994) 'Corporate reporting for sustainable development', *Environmental Values* 3, 1: 190–210.

– (1996), 'Corporate reporting for sustainable development: accounting for sustainability in AD 2000', in R. Welford & R. Starkey (eds), *The Earthscan Reader in Business and the Environment*, London: Earthscan.

Hajer, M. A. (1995) *The Politics of Environmental Discourse: Ecological Modernisation and the Policy Process*, Oxford: Oxford University Press.

Hawken, P. (1994) *The Ecology of Commerce: How Business Can Save the Planet*, London: Weidenfeld and Nicolson.

Hayward, J. E. S. and Watson, M. M. (1975) *Planning, Politics and Public Policy*, Cambridge: Cambridge University Press.

Hillary, R. (ed.) (1997) *Environmental Management Systems and Cleaner Products*, Chichester: Wiley.

Hirst, P. and Thompson, G. (1996) *Globalisation in Question*, Cambridge: Polity Press.

Hunt, D. and Johnson, C. (1995) *Environmental Management Systems*, New York: McGraw-Hill.

Ikwue, A. and Skea, J. (1996) 'Business and the genesis of the European Community carbon tax proposal', in R. Welford and R. Starkey (eds), *Business and the Environment*, London: Earthscan.

Ingelhart, R. (1977) *The Silent Revolution: Changing Values and Politics Styles Among Western Publics*, Princeton, NJ: Princeton University Press.

Irwin, A. and Hooper, P. (1992) 'Clean technology, successful innovation, and the greening of industry: a case-study analysis', *Business Strategy and the Environment* 1, 2: 1–11.

IUCN (International Union for the Conservation of Nature) (1980) *The World Conservation Strategy: Living Resource Conservation for Sustainable Development*, IUCN: Geneva.

Jackson, T. (ed.) (1993) *Clean Production Strategies*, Boca Raton, FL: Lewis Publishers.

Jänicke, M. (1985) *Preventative Environmental Policy as Ecological Modernisation and Structural Policy, Paper 85/2*, Berlin: Wissenschaftszentrum.

Katzenstein, P.J. (1985), *Small States in World Markets*, Ithaca: Cornell University Press.

Kirby, J. (1995) *The Earthscan Reader in Sustainable Development*, London: Earthscan.

Kluppel, H.-J. (1995) 'Life Cycle Assessment with pollution alcohol sulphate', *INFORM* 6, 6: 647–57.

Knox, P. L. and Taylor, P. J. (eds) (1995) *World Cities in a World System*, Cambridge: Cambridge University Press.

Lafferty, W. M. and Eckerberg, K. (1998) *From the Earth Summit to Local Agenda 21: Working towards Sustainable Development*, London: Earthscan.

Lindblom, C. (1977) *Politics and Markets*, New York: Basic Books.

Marstrander, R. (1994) 'Industrial ecology: a practical framework for environmental management', in R. Marstrander (ed.) *Environmental Management Handbook*, London: Pitman.

McCormick, J. (1991) *British Politics and the Environment*, London: Earthscan.

Mazey, S. P. and Richardson, J. J. (1993) *Lobbying in the EC*, Oxford: Oxford University Press.

Meadows, D. L., Randers, J. and Behrens, W. W. (1972) *The Limits to Growth: a Report on the Club of Rome's Project on the Predicament of Mankind, 1978 edn.*, London: Pan.

Mol, A. P. J. (1995) *The Refinement of Production: Ecological Modernisation Theory and the Chemical Industry*, Utrecht: Van Arkel.

OECD (Organisation for Economic Co-operation and Development) (1985) *Environment and Economics: the Results of the International Conference on Environment and Economics*, Paris: OECD.

—— (1989) *The Application of Economic Instruments for Environmental Protection*, Paris: OECD.

O'Riordan, T. and Jordan, A. (1995) 'The precautionary principle in contemporary environmental policy', *Environmental Values* 4, 2: 191–212.

Owen, D. L. (1992) *Green Reporting: the Challenge of the Nineties*, London: Chapman and Hall.

Owens, S. (1997), 'Interpreting sustainable development: the case of land-use planning', in M. Jacobs (ed), *Greening The Millenium?: The New Politics of the Environment*, Oxford: Blackwells.

Pearce, D., Markandya, A. and Barbier, E. (1989) *Blueprint for a Green Economy*, London: Earthscan.

Porter, M. and van der Linde, C. (1995) 'Green and competitive: ending the stalemate', *Harvard Business Review* 73, 5: 120–33.

Redclift, M. (1993) 'Sustainable development: needs, values, rights', *Environmental Values* 2, 1: 3–20.

Reid, D. (1995) *Sustainable Development: an Introductory Guide*, London: Earthscan.

Richardson, D. and Rootes C. (1995) *The Green Challenge*, London: Routledge.

Roberts, P. (1995) *Environmentally Sustainable Business: a Local and Regional Perspective*, London: Paul Chapman Publishing.

Robertson, R. (1992) *Globalisation: Social Theory and Global Culture*, London: Sage.

Rowell, A. (1996) *Green Backlash: Global Subversion of the Environmental Movement*, London: Routledge.

Rudig, W. (1988) 'Peace and ecology movements in Western Europe', *West European Politics* 11, 1: 26–39.

Salt, J. E. (1997) *Catastrophic Risk: an Insurance Perspective*, Boreham Wood: UK Loss Prevention Council.

—— (1998) 'Kyoto and the insurance industry: an insider's perspective', *Environmental Politics* 7, 2: 160–5.

Sassen, S. (1994) *Cities in a World Economy*, Thousand Oaks, CA: Pine Forge/Sage.

Schmidheiny, S. (1992) *Changing Course: a Global Business Perspective on Development and the Environment*, Cambridge, MA: Massachusetts Institute of Technology.

Schnaiberg, A. (1997) 'Sustainable development and the treadmill of production', in S. Baker, M. Kousis, D. Richardson and S. C. Young (eds) *The Politics of Sustainable Development*, London: Routledge.

Selman, P. (1996) *Local Sustainability: Managing and Planning Ecologically Sound Places*, London: Paul Chapman Publishing.

Simonis, U. E. (1989) 'Ecological modernisation of industrial society', *International Social Science Journal* 121: 347–61.

Spaargaren, G. and Mol, A. P. J. (1992) 'Sociology, environment and modernity: ecological modernisation as a theory of social change', *Society and Natural Resources* 5: 323–44.

Stead, W. E. and Stead, J. G. (1996) 'Strategic management for a small planet', in R. Welford and R. Starkey (eds) *Business and the Environment*, London: Earthscan.

Taylor, B. (1991) 'The religion and politics of Earth First!', *The Ecologist* 21, 6: 258–62.

Tibbs, H. (1993) *Industrial Ecology: an Environmental Agenda for Industry*, Emeryville, CA.: Global Business Network.

Vig, N. J. and Kraft, M. E. (1990) *Environmental Policy in Europe and Japan*, Washington, DC: Congressional Quarterly Press.

Walley, N. and Whitehead, B. (1994) 'It's not easy being green', *Harvard Business Review* May/June: 46–52.

Wathern, P. (ed.) (1988) *Environmental Impact Assessment*, London: Unwin Hyman.

WCED (World Commission on Environment and Development) (1987) *Our Common Future*, Oxford: Oxford University Press.

Weale, A. (1992) *The New Politics of Pollution*, Manchester: Manchester University Press.

Welford, R. (1994) *Cases in Environmental Management and Business Strategy*, London: Pitman.

—— (1995) *Environmental Strategy and Sustainable Development*, London: Routledge.

—— (1996) 'Breaking the link between quality and the environment: auditing for sustainability and life cycle assessment', in R. Welford and R. Starkey (eds), *The Earthscan Reader in Business and the Environment*, London: Earthscan.

Welford, R. and Starkey, R. (1996) *The Earthscan Reader in Business and the Environment*, London: Earthscan.

Weston, J. (1989) *The FoE Experience: the Development of an Environmental Pressure Group*, *Working Paper 116*, Oxford: Oxford Polytechnic School of Planning.

Williamson, P. J. (1989) *Corporatism in Perspective*, London: Sage.

Willums, J. and Goluke, U. (1992) *From Ideas To Action: Business and Sustainable Development*, Oslo: Ad Notam Gyldendal.

Wilson, G. A. and Bryant, R. L. (1997) *Environmental Management: New Directions for the 21st Century*, London: UCL Press.

Wintle, M. and Reeve, R. (eds) (1994) *Rhetoric and Reality in Environmental Policy: The Case of the Netherlands in Comparison with Britain*, Brighton: Avebury Press.

Worcester, R. (1997) 'Public opinion and the environment', in M. Jacobs (ed.) *Greening the Millennium: the New Politics of the Environment*, Oxford: Blackwell.

Wood, C. (1995) *Trading in Futures*, Lincoln: Royal Society for Nature Conservation.

Young, S. C. (1996) *Promoting Participation and Community-Based Partnerships in the Context of Local Agenda 21: a Report for Practitioners*, Manchester: Government Department, Manchester University.

Young, S. C. with Lowe, A. V. (1974) *Intervention in the Mixed Economy*, London: Croom Helm.

1   Ecotaxes on the Belgian agenda, 1992-5 and beyond

Environment and economy at the heart of the power struggle[1]

Benoît Rihoux

The ecotaxes, which were brought onto the agenda by the green parties *Ecolo* and *Agalev*[2] in an exceptional political conjuncture, developed into one of the most controversial issues in Belgium from mid-1992 onwards. This chapter starts with a condensed account of this issue within a broader context. The second section analyses the goals, strategies and resources of the main actors involved; while the third section centres on the main features of the negotiation process.

In order to explain why an issue of apparently secondary importance raised so much controversy, we will analyse the wider implications of this particular policy instrument. We argue that it reveals broader issues, such as the legitimacy of the regulatory state's intervention in the economic sphere and the goals and direction of future environmental policies. Finally, we speculate on the prospects and possible impact and legacy of the ecotaxes once the policy window, which allowed their emergence, was shut.

The process: a short account[3]

Historical background[4]

The ecotaxes discussed in Belgium from 1992 onwards can be traced back to *Ecolo*'s 1985 electoral programme. Though these early proposals remained largely unnoticed, they were part of a broader process which initiated in the 1970s at world, European and Belgian levels the entry into politics of environment-related issues and the gradual emergence of economic instruments in environmental policies.

The first environmental policies were almost exclusively centred on command-and-control methods. From the early 1980s onwards, as the limitations of this approach became obvious, more and more environmental economists started to advocate the introduction of economic instruments. These suggestions were not rapidly translated into substantial policies, although the rise of global environmental issues from the mid-1980s onwards did encourage some concrete proposals.

At the European level, one specific economic instrument began to attract more attention from 1989 onwards: the carbon tax. In October 1991 the Commission issued its first proposal in the context of a global strategy; followed by a more precise proposal, which included a conditional clause in June 1992.[5] It was discussed twice by the Council of Ministers in 1993, but no agreement was reached.

At the Belgian level, the entry of environmental issues into politics has been a slow and painstaking process initiated in the early 1970s (Gobin 1986). In 1980 and in 1988, following the federalisation process, the environment became a predominantly regional responsibility, although a few responsibilities remained at the national level. It is thus at the regional level – Flanders, Wallonia and Brussels – that the main environmental policies have been initiated from the early 1980s onwards.[6] However, due to the complexity of the institutional framework, most policies which also have national implications (which includes most environmental policies) usually require lengthy negotiations between the four partners (i.e. the three regional authorities and the federal state). In the subsector of waste management (specifically packaging wastes) which stood at the centre of the debate on ecotaxes, all three regions had initially opted for a *covenant* approach in the late 1980s and early 1990s (i.e. voluntary agreements between the regional executive and economic and industrial federations or individual companies). Essentially, this centred on the development of recycling systems.

The European carbon tax was also discussed in Belgium from 1990 onwards, although it only reached the agenda in April 1992. In 1990 and in 1992 the national *Bureau du Plan* examined various scenarios as well as the impact of such a tax in Belgium.[7] Before the June 1992 Rio summit the national *Conseil Superieur des Finances* backed the proposal. However, the main political parties remained divided on this matter: both liberal parties – *Parti Réformateur Libéral* (PRL) and *Partij voor Vrijheid en Vooruitgang* (PVV) – rejected it, whereas the socialists – *Parti Socialiste* (PS) and *Socialistische Partij* (SP) – and the Christian democrats – *Parti Social-Chrétien* (PSC) and *Christelijke Volkspartij* (CVP)[8] – approved it. The greens, however, advocated more ambitious scenarios. Although the debate continued after 1992, this specific proposal eventually dropped off the agenda.

From the November 1991 elections to the Community to Community Dialogue (April–July 1992)

The general election of 24 November 1991 produced quite a few changes, with notably more MPs for the greens. The four parties of the first Christian-Socialist Dehaene cabinet, finally formed in late February 1992, persuaded both the PRL and the PVV, the ethnoregionalist parties *Front Démocratique des Francophones* (FDF) and *Volksunie* (VU), and the greens to join the Community to Community Dialogue. This strategy was designed to achieve a two-third majority in both Houses, which was needed if the final stage of the institutional reforms transforming Belgium into a federal state was to be passed.[9]

The greens responded rapidly by putting forward ecotaxes as one of their main preconditions for supporting the two-thirds majority. Two types of ecotaxes were therefore formally discussed in one of the Dialogue's working groups: a carbon tax and a variety of product ecotaxes. In mid-July 1992 the Dialogue ended in failure, leaving the four majority parties – PS, SP, PSC and CVP – with the potential support of only three additional parties: the VU and the two green parties. The governmental parties hence agreed to include some ecotaxes in the negotiation platform, in parallel with the institutional aspects. The national carbon tax was, however, dropped and only four categories of product ecotaxes were accepted in principle.

The negotiations (October 1992–January 1993)

Negotiations on the details of the ecotaxes started in early October 1992, following the first proposal produced by the greens concerning seven broad categories of goods (see Table 1.1). An ad hoc working group, headed by an *Ecolo* MP and comprising negotiators from the seven parties, was installed in order to settle this specific issue. From the very start severe disagreements developed within the working group because economic interest groups swiftly produced spectacular reports on the potential negative consequences of the ecotaxes in micro- and macroeconomic terms. The industrial federations, the PRL and the PVV also argued that ecotaxes would endanger the existing regional covenants. A compromise solution was sought by inviting a few representatives from the economic interest groups to working group meetings and by relying on the Prime Minister to act as both a facilitator and an arbitrator.

On 15 November 1992 a first partial agreement was reached within the working group. However, some ecotaxes remained more problematic, especially on some non-reusable drinks containers. This led to more negotiations and to a widening of the working group to include representatives of the three regions and of two national ministries. Whenever negotiations became deadlocked, the Prime Minister drew together the four governmental party presidents in order to find ways of resuming the negotiation process.

A compromise agreement was reached on 8 December 1992 (see Table 1.1). It included the creation of a follow-up committee and an ad hoc advisory board composed of experts designated by the seven negotiating parties. They were to assess the potential economic impact of the ecotaxes and suggest possible adaptations. This first agreement raised much controversy, especially regarding the ecotax on polyvinyl-chloride (PVC) drinks containers. This led some majority parties (especially the PS) to resume the negotiation process. A new compromise was reached on 15 January 1993. It involved some minor modifications, including a widening of the follow-up committee's prerogative. Each type of ecotax was given a precise implementation date between January 1994 and January 1997 (see Table 1.1).

The debate in parliament

In early February 1993 the ecotaxes bill was submitted to parliament as a chapter

Table 1.1 Changes to Belgian ecotax proposals, September1992–April 1995.

	The Greens' initial proposal, 30 September 1992	Agreements, 8 December 1992 and 15 January 1993	Bill, 16 July 1993	Royal Orders, December 1993	Bill, 3 June 1994	Bills, 9 February 1995 and 4 April 1995
1. Drink containers	Rate: 20 BEF/litre; reduced rate (10 BEF/litre) if recycling conditions are met; exemption if full recycling or reuse; implementation: 1 January 1993 (beer containers) or 1 September 1993 (others)	Rate: 15 BEF/litre; only four types of drink containers (soda, beer, sparkling water); reduced rates suppressed; exemption if recycling % are progressively met from 1993 to 1997 (no exemption for PVC containers)	Implementation 1 April 1994 (no modification for PVC containers)		General exemption until 31 December 1994 (condition: the producers must prove that they are making the necessary investments); PVC containers: undefined implementation date (3 months after a report produced by the follow-up committee)	General exemption until 30 June 1995 (so that the follow-up committee can produce a report), then further exemption until 31 December 1995
2. Package of some industrial goods	Rate: 50 BEF/litre; 21 categories of industrial goods; reduced rate (20 BEF/litre) if non-toxicity requirements are met; exemption if stronger reuse or recycling	Rate: min. 25 BEF/litre; only 5 categories of industrial goods (ink, glue, oil, solvents, pesticides); further exemption if 'sufficiently' reused and treated		Implementation: 3 months after the publication of a Royal Order defining the meaning of 'non-professional use' (following a report by the follow-up committee)		

	requirements are met; implementation: 1 January 1994				
3. Other packaging materials	Rate: 20–200 BEF/ kilo; reduced rates if reuse or recycling requirements are met; exemption if stronger reuse or recycling requirements are met; implementation: 1 January 1994	Dropped			
4. One-use products	Rate: from 10 BEF/ unit (razors) to 300 BEF/unit (cameras); 7 categories of products; reduced rates if recycling requirements are met; implementation: 1 July 1993	Only 2 categories of products: razors and cameras; softer recycling requirements for cameras; exemption if reuse requirements are met; implementation: 6 months after the Bill comes into effect	Implementation: 31 January 1994 (razors) and 1 July 1994 (cameras); 31 January 1994: implementation of the ecotax on razors	1 July 1994: implementation of the ecotax on cameras	Cameras: exemption (instead of reduced rate) if reuse or recycling requirements are met
5. Batteries	Rate: 10 BEF/unit (plus additional rate according to chemical	Rate: 10 BEF/unit; - implementation: 1 January 1994	Implementation: 1 July 1994	Implementation: 1 January 1995	Implementation: 1 July 1995, then 1 January 1996; softer exemption rules

Table 1.1 (continued)

	The Green's initial proposal, 30 September 1992	Agreements, 8 December 1992 and 15 January 1993	Bill, 16 July 1993	Royal Orders, December 1993	Bill, 3 June 1994	Bills, 9 February 1995 and 4 April 1995
	composition); exemption if deposit system; implementation: 1 July 1993					General exemption for 1 year (until 31 December 1995)
6. Paper and cardboard products	Rate: 10 BEF/kilo; reduced rate if recycling and non-toxicity requirements are met; implementation: 1 July 1993	Further reduced rates if no chlorine used; softer recycling rates for exemption (to be met progressively from 1994 to 1999); implementation: 1 January 1994–1 January 1996 (according to the type of paper products)	Softer recycling rate for exemption (newspaper paper); implementation: 1 year later (1 January 1995) for newspaper paper	General exemption for 1 year (until 31 December 1994)		
7. Pesticides and phyto-sanitary products	Rate: 2–10 BEF/g of active substance (according to level of toxicity); reduced rate for professional use and products accepted in	Exemption for professional users, 'less harmful' substances which have no substitute in biological agriculture,	List of products can be modified; exemption (until 31 December 1994) for professional treatment of paper and cardboard	Implementation: 1 April 1994 (if the follow-up committee produces a report in time)	Implementation: 3 months after the list of products has been adapted (following a proposal by the follow-up committee)	

'biological agriculture'; implementation: 1 January 1993	professional wood treatment products (until 31 December 1994) and some non-agricultural products used as disinfectants; implementation: 6 months after the Bill comes into effect	products; implementation: 31 January 1994

Sources: Rihoux (1994b: 66–83) and CEFE (1995).
Note: Only the main modifications are mentioned. See sources for detailed information.

of the bill on institutional reforms. In mid-March 1993 an alternative bill was presented by the FDF. It advocated the introduction of ecofees: low-level fees which would be paid by the producers – per unit of product – in order to finance waste-collecting and -recycling systems. This was midway between the German *Duales System* and the French *Eco-emballages* in terms of scope and ambition. Similarly, several economic federations advocated a broadening of the existing covenants, as well as a more modest ecofee system, as an alternative to the ecotaxes.

The ecotaxes and ecofees proposals were discussed jointly in the chamber (May 1993) and in the senate (July 1993). The debate centred mainly on the potential economic impact of the ecotaxes, on the PVC drinks containers case, on the implementation details and on the (in)compatibility with the existing covenants.[10] In the meantime (April 1993), the three heads of the regional executives and their environment ministers signed a cooperation agreement regarding the partition and allocation of the ecotaxes' receipt. It was to be directed solely towards regional environmental rehabilitation programmes. The bill on the federalisation of Belgium – including its ecotaxes chapter, with only minor modifications (see Table 1.1) – was eventually passed in the chamber (11 June 1993) and the senate (14 July 1993). The ecofees were rejected.

Towards implementation?

The follow-up committee was only established in late November 1993, 3 months behind schedule. Its first task was to examine the PVC drinks-containers question and some practical details regarding the implementation of the first ecotaxes. Pressed by the Prime Minister, the committee refused to answer the question as to whether the implementation timetable should be partly or totally modified.

On 21 December 1993, despite protests from the greens, the federal government announced modifications in the implementation timetable of some ecotaxes (see Table 1.1) and asked the follow-up committee to examine other ecotaxes quickly as well. Two days later, blaming insoluble technical problems,[11] the government issued a 1-year full exemption for paper products (see Table 1.1). On 31 January 1994, following unsuccessful legal procedures introduced by several producers, the first ecotax – on disposable razors – came into effect. In the meantime, the follow-up committee painstakingly started to examine the case of PVC drinks containers.

On 3 June 1994 the four majority parties once again introduced modifications to the timetable, for all types of ecotax this time; from a few months to an undefined date depending upon the progress of the follow-up committee's work (see Table 1.1). The committee thus received additional requests from the government. On 1 July 1994 the ecotax on disposable cameras came into effect.

On 9 February 1995 the four majority parties introduced delays for most of the ecotaxes which were not yet being implemented, including a further 1-year full exemption for paper and cardboard products (see Table 1.1). In April 1995, after the announcement of anticipated general elections on 21 May, the elaboration of the official information campaign on ecotaxes came to a halt. By that time the follow-up committee had only produced some of the required reports and the excise administration was still a long way off completing the necessary preparatory work for the implementation of the remaining ecotaxes. On 4 April 1995 further delays were decided on for drink containers and batteries (see Table 1.1). At that point the implementation date for some product ecotaxes was announced by the government for early 1996, while other ecotaxes (package of industrial goods, pesticides and paper products) were still awaiting the reports from the follow-up committee and further administrative work.

Actors around a controversial issue: aims, strategies and resources

The green parties

From a time perspective, the ecotaxes put forward by the green parties in 1992 have undoubtedly not been implemented overnight. On the one hand, they can be traced back to 1985. In this respect one observes that *Ecolo* has placed a more precocious and stronger emphasis on them than *Agalev*. On the other hand, the greens in the European Parliament had already been strong advocates of the European carbon tax from 1989 onwards. Interestingly, the leading green (*Ecolo*) MP on the issue worked as an expert on this very subject for the greens in the European Parliament from 1989 to 1991. The product ecotaxes came more as a surprise to the other political parties. This probably proved profitable for the greens – at least during the first round of negotiations – since it took some time for the other parties to mobilise enough expertise.

Moreover, *Ecolo* and *Agalev* were generally able to remain united throughout the negotiation process. This was made possible by the existence of a joint parliamentary group as well as other structural and more informal links between the two parties. The main source of disagreement was the allocation of the receipt of the carbon tax. The Flemish parties – including *Agalev* – feared that it would be partly diverted from environmental programmes to the education sector under the pressure of the French-speaking parties.[12]

The internal consensus level within each green party was not there either. Participating in the Community to Community Dialogue and the fact that they had to accept successive compromise solutions were quite controversial in both parties. This has to do with the fact that the grass-roots party activists are less participationist than the party professionals (staff and MPs). To a certain extent the latter have integrated some elements of the established Belgian political culture and practices, especially negotiating procedures and compromise agreements. In both parties this generated quite a few tensions; not only between the professionals and the rank-and-file activists, but also between different segments of the party elite: mainly the negotiators and MPs on the one hand, versus the party executive and party council on the other.[13] These tensions are symptoms of the structural stress experienced by the green parties after the 1991 elections due to their increasing size.

On the other hand, a striking feature of the ecotaxes debate is the weak presence of environmental organisations, both in the mass media and in terms of proposals and reactions. This is not so surprising, as one should not over-estimate the resources and influence of the Belgian environmental movement as compared with neighbouring countries, e.g. the UK and the Netherlands (Walgrave 1994; Rihoux 1995a). Furthermore, their impact on policy-making is undoubtedly much weaker than that of the industrial federations and trade unions, because most environmental organisations are quite recent and very diverse in terms of size, stability and aims (Walgrave 1994). More importantly, with a few exceptions (Rihoux 1995a; Walgrave 1995: 55), they are virtually absent from the Belgian institutionalised, neocorporatist concertation mechanisms.[14] Secondly, the green parties did not – or were not able to – mobilise these organisations due to the absence of structural links (Kitschelt 1989; Kitschelt and Hellemans 1990; Rihoux 1995b; Walgrave 1995). The dominant strategy of these movements involves political pluralism, a search for positive contacts, lobbying the governmental parties, a quest for respectability, and little adversary politics (Rihoux 1995a; Walgrave 1994,1995). Thirdly, most environmental organisations did not feel they had to react strongly on the ecotaxes, either because they were not convinced of their opportuneness or because, for once, they were not in a defensive position. They did not feel they had anything to lose in the process.

However, some environmental organisations became more mobilised in connection with the polarisation on the PVC ecotax. In fact, the two most active ones – Greenpeace Belgium and the *Bond Beter Leefmilieu*, a Flemish umbrella organisation – had already campaigned against PVC and chlorine production in the early 1990s (Clement *et al.* 1993).

How can one then provisionally assess the relative success or failure of the greens on ecotaxes? If one concentrates on the ecotaxes alone then the picture is quite bleak. On the other hand, they brought a new issue onto the agenda and managed to translate some proposals into a bill backed by governmental parties. However, the concrete implementation of some of the product ecotaxes became increasingly uncertain. Indeed, the follow-up committee's resources were insufficient to cope with its increasing responsibilities and the agreed-upon timetable was modified considerably four times by the majority parties. Moreover, the greens could not get agreement on the carbon tax, which would assuredly have had a much more wide-ranging impact in environmental and economic terms. Finally, they had to drop many of their initial proposals (see Table 1.1).

The other political parties

It is necessary to identify the five parties which joined the greens in the ecotax negotiations. These were the four majority parties – the PS, SP, PSC and CVP – and the VU. The others were the outsiders: the PRL, PVV/VLD[15] and FDF.[16]

It looks as though each one of the majority parties has sought to maximise its political benefits (the greens' support to the institutional reforms) while minimising its political costs (the ecotaxes). Indeed, they have tried to push the greens as close as possible to their breaking point over the ecotaxes. In addition, relying on classical negotiation mechanisms, they tried to make sure that the greens became embedded in all the parallel negotiation processes: institutional, financial and ecotaxes. Hence, if the greens left the negotiation table they would bear responsibility for the failure of the whole process.

On the other hand, some conflicts which occurred during the negotiations can be partly explained by several of the constraints faced by some of the majority-party leaders (notably ministers). Firstly, the implementation of ecotaxes was bound to have some practical, and disturbing, consequences for waste-management policies (particularly the existing regional covenants), fiscal policies and the reform of the fiscal administration (as the ecotaxes would necessitate major changes in customs' administration) and economic policies (particularly the subsidising of industrial activities).[17] Secondly, all majority-party leaders needed to avoid alienating their electoral bases and intermediate organisations within their own pillar.[18] For instance, the rejection of the PVC ecotax by the PS president was caused by the strong reaction of the socialist trade unions' grass roots in some chemical companies. Finally, the strategists of the majority parties (especially the PS) attempted to use the ecotaxes discussion to affect the credibility of the greens on socio-economic issues by picturing them as pro-environment and anti-employment.

From the very start, the PRL and the PVV/VLD uncompromisingly opposed any kind of ecotaxation, because they positioned themselves as the defenders of the economic sphere and private initiative against the growing (fiscal) pressure of the regulatory state. In contrast, by advocating ecofees as an alternative to the product ecotaxes, the FDF[19] tried to position itself as a party of realist environmentalists against the utopian greens and as a responsible actor in the framework of existing environmental policy. However, with only a few MPs, the FDF was not able to influence decisions.

The economic interest groups and trade unions

At first, the economic interest groups (i.e. mainly the industrial federations and their umbrella organisations) found it hard to take the issue seriously as the outcome of the Community to Community Dialogue seemed uncertain. However, some industrial groups were cautious enough to try to accelerate the signing of the waste-management covenants in Brussels and Wallonia, in order to claim the existence of a viable alternative to possible ecotaxes. During a second period the business community mobilised almost unanimously against the ecotaxes proposals, mostly stressing their negative macro- and microeconomic consequences. The third phase gradually developed as soon as it became clear that the ecotaxes working group would reach some sort of agreement. The strategy of the pressure groups then became more subtle. On the one hand, they tried to extend their pressure beyond the media by lobbying the ministers and party leaders and by claiming a place at the negotiating table. On the other hand, their critiques became more elaborate as they questioned their ecological appropriateness and stressed some scientific uncertainties. From then on one observes a dualisation of their discourse: uncompromising opposition towards the media, but a compromise-seeking approach towards the political negotiators. Finally, after the January 1993 agreement, the pressure groups adopted an even more elaborate strategy by threatening legal actions; by asking to be associated with the follow-up on ecotaxes; and by attempting to produce an alternative by advocating the introduction of more ambitious covenants as well as some (modest) ecofees.

These four stages are not as clear-cut as they seem, however, for one observes little real unity in the economic interest-groups' moves (Clement *et al.* 1993: 143–4). This is quite understandable considering the differing interests and strategies of the various companies and sectors in a competitive market economy. On several occasions the representatives of the industrial sectors pursued their own company's specific goals in order to try and reinforce their market position against their competitors (De Clerq 1994: 12). Furthermore, in most sectors concerned the interests and strategies of the producers and distributors often differed significantly.

The initial – rather favourable – reactions of the two major (Christian and socialist) trade unions at the national level changed rapidly under the influence of at least three factors: the centring of the debate around the employment issue; the vigorous reaction and media campaigning of most industrial federations; and – last but not least – the very strong negative backlash of the trade union grass roots, mostly in chemical plants dealing with PVC and plastics production.

Initially, this eventually led to an unusually high[20] level of consensus and cooperation, such as jointly organised anti-ecotaxes demonstrations between the trade unions and the employers' organisations. Quite rapidly though, the trade unions chose to distance themselves from the industrial federations due to the growing number of employer–trade union conflicts linked with severe difficulties in numerous factories and sectors. The unions feared that the ecotaxes would be used by the employers to justify transfers of factories or businesses to other areas, although the decision would have been made on other grounds. Throughout the discussions the trade unions remained in an uncomfortable position because they had to rely on figures and analyses provided by the employers' organisations in order to assess the impact of the ecotaxes.

All the actors have thus had to adapt their strategies according to the interaction and negotiation patterns, the ever-changing balances of power and the limits of their internal resources. Furthermore, each organisation has been more or less divided as to the best strategy to be followed, as internal elite coalitions formed and collapsed according to the variations of the internal balance of power and external pressures.

The negotiation process: time, procedures and policy entrepreneurs

Battling around the time factor

The process of the ecotaxes negotiation was divided between detailed discussions and marathon negotiation sessions leading to partial agreements. During each one of these phases, as negotiations became blocked within the working group, bilateral contacts – most often initiated by the Prime Minister – took place and eventually led to a compromise solution accepted by the seven party presidents or representatives.[21] Immediately before, during and right after each discussion phase both the economic and social interest groups, the liberal parties and the greens strove to increase their pressure on the majority parties through various means. In this process the influential daily newspapers played a partisan role, as most newspapers are tied to a particular pillar.[22]

Furthermore, the negotiation timetable gradually became a negotiation issue in its own right. For instance, the successive final deadlines, which were changed at least four times, for the vote in parliament were designed by the Prime Minister to avoid an overlap between the ecotax negotiations and the other institutional and financial negotiations together with the budgetary discussions within the government. They were also used – both by the Prime Minister and by the greens – to try and force a final agreement.

Negotiation mechanisms and procedures

The negotiation process on ecotaxes was characterised by pragmatism, informality and inventiveness. Both for the ecotaxes and for the institutional reforms, the negotiators – the Prime Minister especially – have resorted to a variety of techniques associated with the usual consociational, Belgian-style elite-accommodation mechanisms.[23] From the very start of the Community to Community Dialogue, the complex and conflict-laden issues were sliced up into more readily negotiable sub-issues shared out among several working groups. This minimised the risk of general confrontation between the negotiating partners and made it increasingly difficult for one partner – the greens especially – to leave the negotiation table. The later forming of the ad hoc ecotaxes working group only prolonged this technique. Furthermore, the negotiators started with the agreed-upon points and left the most conflict-laden issues aside for the final phase (Blaise and Lentzen 1993: 53). Once negotiations became stranded they proceeded with give-and-take compromise agreements or even divided the matter once again into even smaller units. Eventually, the ecotaxes and the institutional reforms were aggregated into one large omnibus bill. On the whole these various techniques have clearly been more beneficial to the majority parties than to the greens, because the latter have been forced to retreat after each slicing up of the ecotaxes. In fact, their major retreat was to accept the withdrawal of the carbon tax.

On the other hand, the ecotaxes working group had a predominantly informal and at times ambiguous character. Its composition changed through time and its members' mandate was often ambiguous, as some of them were both political negotiators and experts. Moreover, their discussions were regularly bypassed by the parties and by the Prime Minister. To a certain extent, some conflicts can be better explained by these ambiguities than by substantial disagreements on the content of the ecotax proposals. Such conflicts are bound to carry on since the follow-up committee's task and composition retained the ambiguities.

Putting ecotaxes onto the political agenda

Why did the greens choose the ecotaxes as their part of the bargain in return for giving their support to the two-thirds majority on the constitutional issues? In the process, how did they manage to bring this controversial issue, which nobody else seemed to want, onto the legislative agenda? The first question leaves room for much speculation. To start with, it is not exactly clear that the greens' initiative was so rational.[24] To a certain extent, one could contend that the greens needed to put forward something as a bargaining counter. However, there are some indications that the greens chose this particular issue in order to try and raise a series of fundamental questions and promote changes in environmental policies, especially covenants. Hence, there was probably some sort of – arguably over-ambitious – strategic rationale behind the ecotaxes proposal.

The most obvious answer to the second question is that the greens were simply able to seize a unique – albeit ephemeral – opportunity. Drawing from two complementary theoretical models we would like to push this argument a bit further. Wilson (1980): 366) argues that 'policy proposals, especially those involving economic stakes, can be classified in terms of the perceived distribution of their costs and benefits'. Wilson contends that one expects to find what he calls 'entrepreneurial politics' when a policy proposal induces widely distributed benefits and narrowly concentrated costs. Such policy proposals require 'the effort of a skilled entrepreneur who can mobilise latent public sentiment [...], put the opponents [...] publicly on the defensive [...] and associate the legislation with widely shared values' (Wilson 1980: 370). He further argues that this is where most environmental policies stand. Indeed, the Belgian product ecotaxes are bound to have direct or indirect structural impacts on some specific industrial and commercial sectors; these are the concentrated costs. They also aim at achieving better environmental conditions for all individuals; which is spreading benefits.

In this case, the greens collectively have acted as a policy entrepreneur, mobilising predominantly political resources, and exploiting a temporarily open policy window in a peculiar political conjuncture. Kingdon's model, inspired notably by Cohen, March and Olsen's garbage can model, seems quite relevant in this respect (Kingdon 1984; Cohen *et al.* 1972). According to this model, three major streams of processes run through the governmental machine: problems, policies and politics. At certain critical times – the policy windows – these separate streams come together: 'a problem is recognised, a solution is available, the political climate makes the time right for change, and the constraints do not prohibit action' (Kingdon 1984: 93). Two points in Kingdon's argument must be stressed: the main sparkle usually comes from the politics stream. This is the change in the political conjuncture. Also the coupling of these three streams requires the intervention of an entrepreneur. This is exactly what happened with the ecotaxes: environmental problems (for example waste-management problems) had been on the agenda for two decades, ecotaxes proposals had been floating around for some time, and at one point a temporary political opportunity became available. The greens were the 'entrepreneurs' who were able to take advantage of it.

Kingdon also observes that interest groups most often try to affect the governmental agenda 'more by blocking potential items than by promoting them' and by trying 'to insert their preferred alternative into a discussion once the agenda is already set' (Kingdon 1984: 71). This is clearly what happened here, because the business community tried to oppose the ecotaxes both in a formal way and by advocating the pursuit of the covenants policy or even the implementation of some sort of ecofee system.

Beyond the policy instrument: some broader implications

Industry's attitude

All things considered and excluding the carbon tax, one could argue that the minor product ecotaxes which were finally included in the 1993 omnibus bill are not so significant in terms of economic impact or structural change. Why then do they still stir so much controversy? How can one account for the apparent disproportion between the vehemence of the negative reactions of the economic interest groups and the, arguably, almost innocuous character of – at least some of – the product ecotaxes? To start with, how significant are the product ecotaxes in terms of their economic and environmental impact? No one seems to agree on this point, which also illustrates a major difficulty: the absence of objective data and analysis on which all the intervening actors could agree (De Clerq 1994: 6– 8).[25]

Some of the product ecotaxes concentrate on a very narrow set of products – disposable razors or cameras, for example. Hence, they only affect a negligible sample out of the large variety of throwaway products and can therefore only be expected to have a marginal – almost symbolic – environmental impact. That was precisely the greens' viewpoint: in a few sectors they selected a limited set of products that could also have a specific value with regard to environmental education. Of course, there would still be a strong direct economic impact from the perspective of the producers and distributors of these categories of products. However, this experience has shown that the producers were able to adapt very rapidly to the new situation by, for example, recycling the components of the disposable camera. This minimised the economic impact of the proposal.

Other product ecotaxes, on drinks containers for example, concern a broader array of products and larger sectors of the economy. Their implementation would certainly bring about significant changes in the production processes and distribution channels, because they would clearly give a competitive advantage to some products over others; for example, reusable over non-reusable polymers. But even in that case, the expected structural impact – as in jobs – is not at all clear. Green experts contend that the ecotaxes will, on the whole, create jobs, whereas the industrial federations anticipate huge job losses. The real conclusion could be that the impact in terms of employment is marginal in most cases, as the follow-up committee has demonstrated in the case of the PVC ecotax.[26] So the question remains: why were the reactions of the business community so fiercely and unanimously negative?

Perhaps the answer can be found in the way the ecotaxes were decided. In this respect one observes that the negotiations took place mainly between the political parties, thereby largely bypassing the neocorporatist mechanisms[27] which are usually predominant in the socio-economic field. Furthermore, trade unions and economic organisations traditionally play a leading role in these mechanisms (Compston 1994: 134–6; Smits 1979) and in some cases take advantage of their preferential links with the political party of their pillar.[28] In this case they were not only confronted with an issue which was discussed outside their realm of influence. They were also faced with a new actor – the green parties – with which they could not establish stable structural links as via ministerial cabinets, party study centres or high-ranking public servants[29] in order to exert pressure. This could explain, to a certain extent, why so many economic actors reacted strongly against the ecotaxes and why they demanded (and obtained) the urgent consultation of the Central Economic Council and other consultative bodies.[30] But this still does not explain the virulence and the apparent quasi-unanimity of these reactions.

Could the explanation then be more of a cultural nature? None of the opponents of ecotaxes actually argued – as some of the predecessors would have done in the 1960s – that environmental protection, on one hand, and economic growth and/or employment, on the other hand, should be opposed in either/or terms. Most economic actors claimed that they wanted to achieve both production goals and better environmental protection, but that the ecotaxes were simply not the right way. One must therefore have a closer look at the essence of ecotaxes as policy instruments.

Policy instruments and policy styles: strong regulatory state or economic laissez-faire?

Let us first look at the definition and the nature of ecotaxes, as compared with the other main instruments of the Belgian national and regional environmental policies. Basically, the Belgian ecotaxes are product ecotaxes with an incentive character: through their high rate their aim is to reduce significantly the production and consumption of environmentally damaging products and to foster the production and consumption of more environment-friendly products. Hence, the product ecotaxes affect, not marginally but substantially, ways of producing products. This certainly constitutes a major break from the prevailing policy instruments.

This should be put into a time perspective. From the first national environmental bills of the 1970s to the national and regional policies of the late 1980s, as in most other countries, Belgian environmental policies have always predominantly followed a command-and-control logic. Then, more recently, while the command-and-control approach remained dominant, a few elements of covenant policies were introduced at the regional level, as mentioned above. According to Andersen's typology of policy designs (see Chapter 4, Figure 4.4) covenants are preventive and consensual, whereas ecotaxes are preventive but coercive.

Hence, if one assumes that future environmental policies are bound to be predominantly preventive, the Belgian product ecotaxes raise the following fundamental question: which kind of preventive instruments will be given priority in the years to come? This is connected with the following underlying question, although never explicitly discussed as such: what should be the role of the regulatory state? Should it follow a predominantly coercive or consensual approach? Our central hypothesis is that the ecotaxes have been (and still are) considered by most intervening actors – whether consciously or unconsciously – as a crucial issue. This is not primarily because of their disputed concrete short-term impact, but because they raise the broader question of the future direction of Belgian federal and regional environmental policies: should they be mainly coercive or predominantly consensual?

Amongst the various intervening actors, the two most polarised positions in this respect are surely occupied by the economic interest groups and the greens. The economic interest groups clearly rule out any coercive solution, be it reactive or preventive. Basically, they oppose any increase in the regulatory powers of the state, especially if it is inspired by political ideology from the greens' perspective. They strongly oppose any attempt of outside authorities to redirect, or even influence, their main production and marketing strategies. What seems to be at stake here is industry's ability to prevent government from changing its conventional approach. It is quite clear that the prevailing regulatory, command-and-control or covenant policies have made it much easier for the business community to influence policy (De Clerq 1994: 8).[31] It is therefore only logical that the business community advocates covenants, possibly financed by some modest ecofees, because these two instruments would not cause any major structural impacts and would not alter the balance of power between government and industry.[32] Conversely, the greens have a much more voluntaristic and political viewpoint. They basically contend that the public authority at whatever level has a legitimate right to actually regulate, to substantially reorient choices to ensure a higher level of environmental quality, for example. This viewpoint legitimises the implementation of policy instruments which could have strong structural impacts even if they are damaging in classical economic terms at the micro- and macro-levels.

To sum up, the main reason why the positions of the advocates and opponents of product ecotaxes have become so polarised is that this particular policy instrument raises a fundamental question: who should have the final say in the regulation of economic activities in a situation in which there is no magic wand which could decisively settle the discussion on the comparative environmental merits of different products and different policy instruments? Who will dominate in the final – inevitable – trade-offs? In this respect the ecotaxes came to be seen as the first battle in a long series of battles: the one that could not be lost.

Conclusion

The years after 1995 proved that, as far as ecotaxes were concerned, the greens had been unsuccessful policy entrepreneurs. They lost their position of being able to 'blackmail' their coalition partners after the last vote on the institutional reforms in July 1993. The policy window closed. Since then, there has been a long story of moves away from the actual implementation of ecotaxes on a significant scale. Only two minor ones were implemented – those on disposable razors and cameras. The more ambitious ones were dismantled, delayed or suspended. Moreover, most producers of batteries and of disposable razors and cameras obtained reduced ecotax rates, or even total exemption.

At the 1994 European elections and then at the May 1995 general elections, both *Ecolo* and *Agalev* suffered bigger electoral defeats than ever before (Rihoux 1995c, 1997). Although they retained a comfortable parliamentary representation, it was clear that their involvement in the whole ecotax affair had a negative impact on their image amongst voters. After the May 1995 election, a Christian-Socialist coalition was confirmed at the federal level with the same prime minister as before; and in most executives at the regional and lower levels of government. As the coalition partners moved slowly but surely away from any serious attempt at implementation of the ecotaxes, the greens gradually withdrew from the monitoring committee between 1995 and 1997. It became clear that the ecotax episode had been a major defeat for *Ecolo* and *Agalev*.

However, it would be wrong to conclude that the ecotax debates had no impact. In several sectors, they have generated or accelerated numerous initiatives aimed at recycling and reuse (CEFE 1995; De Clerq 1994; Defeyt *et al.* 1994). The covenant and ecofee approach was extended. A large specialised agency – Fost Plus – financed by industry and the distribution sector, was created in order to develop a countrywide waste recycling collection and treatment scheme.

Through the 1990s the lesson of this disappointing episode became clear to the greens. In the Belgian context they cannot hope to modify public policies in a substantial way unless they gain a firmer hold on political power and stronger links to the government machine. From 1996 on, learning the main lesson of their political defeat, the greens focused on broader, strategic initiatives. These were partly successful. They established increased contacts, both formal and informal, with the stronger trade unions. Links to these had been monopolised by the Socialist and Christian Democrat parties – the largest unions belong to these two pillars.[33] The greens also revised their socio-economic programme, playing down the role of ecotaxes. Instead they put more emphasis on trying to green the tax system as a whole, by for example proposing 'ecobonuses', or positive incentives to encourage more environmentally friendly travel and consumption patterns. It became clear that one of the lessons of the ecotax episode was simply that the word '*eoctax*' was very unpopular in a country where voters complain regularly about high taxes.

A few weeks before the June 1999 general election, most observers judged that environmental issues had a lower saliency than in the late 1980s and early 1990s. Then the situation was suddenly transformed. A scandal erupted involving large-scale poisoning of the food chain by dioxins. *Ecolo* and *Agalev* seized the opportunity and achieved record electoral gains (Hooghe and Rihoux 2000). As a result they were able to force their way into a 'rainbow' coalition with liberals and socialists at the federal level, as well as in many executives at the regional and lower levels. Learning from experience, they did not try to insert over-ambitious and unrealistic ecotax proposals into the coalition agreement. Instead they set out to try and add green dimensions to fiscal reform programmes – fiscal reform being one of the liberals' top priorities. In addition, the greens aimed to exploit their position to force the debates about the links between the economy and the environment into the open whenever possible. Although they have abandoned their over-ambitious plans for large-scale ecotax reforms, they are still pursuing this goal by trying to promote a package of coercive and preventative policy instruments (see Chapter 4). This is one of the reasons why they held out for relatively non-environmental ministerial portfolios such as transport, energy and public health. Their aim was to have ministers in big spending departments with a variety of links to economic issues. The Pandora's Box that was only opened temporarily in the early 1990s seems likely to be forced open and kept open during the next few years.

Notes

1 The author wishes to thank Martial Mullenders for first-hand information and comments on further developments. A first account was published in Rihoux (1994c).

2 *Ecolo*, the French-speaking green party, covers Wallonia and Brussels; *Agalev*, its Flemish-speaking counterpart, covers Flanders and Brussels.

3 For a comprehensive account of the negotiation process from April 1992 to January 1994, see Rihoux (1994b).

4 For more details, see Rihoux (1994a).

5 The tax would only be implemented if Europe's main competitors – the USA and Japan – introduced similar instruments as well.

6 Belgium was initially a unitary state. In 1993, following a gradual process which was initiated in the 1960s, it became a federal state composed of three communities (the Flemish community, the French community and the German community, mainly dealing with cultural and educational affairs); and three regions (the Flemish region, the Walloon region and the Brussels region). For a comprehensive description of the post-1993 federal system, see Brassinne (1994); Fitzmaur-

ice (1996). For a general presentation of Belgian society and politics, see Fitzmaurice (1988,1996) and Lijphart (1981a).

7 Several economic studies on ecotaxes were conducted in Belgium in the early
   1990s; see the list of sources in Rihoux (1994a).

8 In the course of the 1970s, the three large traditional parties split along the
   linguistic cleavage; see Deschouwer (1994).

9 See note 6. This was a logical follow-on from the 1970, 1980 and 1988 reforms,
   which had left the country in a hybrid quasi-federal situation.

10 On this point some opponents argued that ecotaxes would constitute a break in
   the voluntary approach of the covenants and drastically reduce the production of
   some goods which were already subject to the covenants.

11 The practical impossibility of accurately measuring the percentage of recycled
   fibre in a given paper product.

12 The financing of the education sector remains a major problem for the French
   community.

13 There is no substantive *fundi* versus *realo* polarisation in the Belgian green parties.

14 On neocorporatist decision-making processes in Belgium, see Dewachter (1992:
   127–62), Dewachter and Das (1991: 61–72), Lijphart and Crepaz (1991); Michels
   and Slomp (1990); Smits (1979); Van den Brande (1980);Van den Brande, 1987),
   Van den Bulck (1992); Vanderstraeten (1986).

15 The PVV was transformed into the VLD (*Vlaamse Liberalen en Democraten*) in
   November 1992.

16 This excludes the extreme-right *Vlaams Blok* and *Front National* and the anarchist-
   libertine *Rossem*, which were not invited to the Community to Community Dialogue anyway and did not play a significant role on this issue.

17 In numerous cases the public authority would encourage the development of
   some sectors (particularly heavy industry), while discouraging some activities
   in these same sectors through the implementation of ecotaxes.

18 Belgian society is often defined as pillarised because the largest established
   parties (socialist, Christian democrats, liberals) aggregate, articulate, promote
   and implement the interests of a wider network of organisations which constitute
   their respective pillar. These include trade unions, farmer and middle-class organisations, social and cultural organisations, health-service networks, educational
   and youth organisations, press networks and so on. On pillarisation (or *verzuiling*)
   in Belgium, see Dewachter (1987); Frognier and Collinge (1984); Huyse
   (1971,1981,1987); Lorwin (1974); Pijnenburg (1984); Steininger (1977). There
   is also a growing body of literature on the – disputed – more recent depillarisation
   (*ontzuiling*) processes (Billiet and Dobbelaere 1986; Govaert 1991; Hellemans
   1990; Walgrave 1995).

19 The main initiator was the Europe–Regions–Environment movement, which was
   closely linked with the FDF after the 1989 Brussels regional elections. It was
   founded by a former *Ecolo* MEP, who had left the party following personal and
   financial disputes.

20 At least by Belgian standards, although such cooperation can generally be
   expected when environmental issues are at stake as both employers and unions
   wish to reduce the influence of the authorities on polluting industries (Van der
   Straaten 1992).

21 The greens have no party president.

22 See note 18.

23 On consociationalism in Belgium, see Deschouwer (1994); Lijphart
   (1977,1981b); Lorwin (1974); Pijnenburg (1984) and Steininger (1977).

24 See Kingdon's critique of the rational comprehensive model. In real life, goals are
   not always clearly defined, alternatives are not systematically compared, and the

technology is often unclear: it is more a process of trial and error and pragmatic intervention at critical times (Kingdon 1984: 82–9).

25 In this respect, the ecobalances did not become the magic tool which could settle the discussion once and for all; quite the opposite.

26 The full implementation of the ecotax on PVC drinks containers was only due to affect 1.5 per cent of PVC production in Belgium.

27 See note 14.

28 Especially the trade unions, see note 18.

29 In the case of the large pillarised parties, such links play an important role (De Winter 1989), see note 18.

30 Such as the regional economic councils. There are over a hundred other councils through which employers and unions are consulted on the various aspects of economic policy; see note 23.

31 Besides, business interests have always benefited from the structural and chronic ineffectiveness of the Belgian command-and-control policies, due to insufficient control and a virtual lack of sanctions (De Clerq 1994: 8–9).

32 One could argue that, compared with the command-and-control approach, the covenant approach is even more beneficial to the business community in terms of the balance of power.

33 See Note 18.

References

Billiet, J. and Dobbelaere, K. (1986) 'Naar een desinstitutionalisering van de katholieke zuil?', *Tijdschrift voor Sociologie* 7, 1/2: 129–64.

Blaise, P. and Lentzen, E. (1993) 'La mise en oeuvre des priorités du gouvernement Dehaene. 1. La réforme des institutions', *Courrier Hebdomadaire du C.R.I.S.P.* 1403–4: 52–5.

Brassinne, J. (1994) 'La Belgique fédérale', *Dossier du C.R.I.S.P.* 40: 99–103.

CEFE (Centre d'Etudes et de Formation en Ecologie (Ecolo)) (1995) 'Les écotaxes', unpublished text, Namur: Centre d'Etudes et de Formation en Ecologie (Ecolo).

Clement, J., D'Hondt, H., Van Crombrugge, J. and Vanderveeren, C. (1993) Het Sint Michielsakkoord en zijn achtergronden, Antwerpen, Apeldoorn: MAKLU.

Cohen, M. D., March, J. G. and Olsen, J. P. (1972) 'A garbage can model of organizational choice', *Administrative Science Quarterly* 17, 1: 1–25.

Compston, H. (1994) 'Union participation in economic policy-making in Austria, Switzerland, the Netherlands, Belgium and Ireland, 1970–1992', *West European Politics* 17, 1: 123–45.

De Clerq, M. (1994) 'The political economy of Green taxes. The Belgian experience', *Unpublished Report for the Nordic Council Seminar on 'Economic Instruments in Environmental Policy in a Europe without Border Controls', revised,* Helsinki, 15–16 September.

Defeyt, P., Lannoye, N. and Mullenders, M. (1994) *Ecologie et Fiscalité,* Bruxelles: CED Samson.

Deschouwer, K. (1994) 'The decline of consociationalism and the reluctant modernisation of Belgian mass parties', in R. Katz and P. Mair (eds) *How Parties Organize. Change and Adaptation in Party Organizations in Western Democracies,* London: Sage Publications.

Dewachter, W. (1987) 'Changes in a particratie: the Belgian party system from 1944 to 1986', in H. Daalder (ed.), *Party Systems in Denmark, Austria, Switzerland, The Netherlands and Belgium,* London: Pinter.

— (1992) *Besluitvorming in Politiek België*, Leuven, Amersfoort: Acco.

Dewachter, W. and Das, E. (1991) *Politiek in België: geprofileerde machtsverhoudingen*, Leuven, Amersfoort: Acco.

De Winter, L. (1989) 'Parties and policy in Belgium', *European Journal of Political Research* 17, 6: 707–30.

Fitzmaurice, J. (1988) *The Politics of Belgium. Crisis and Compromise in a Plural Society*, 2nd edition, London: C. Hurst & Co.

— (1996) *Belgium: a Unique Federalism*, London: C. Hurst & Co.

Frognier, A.-P. and Collinge, M. (1984) 'La problématique des "mondes sociologique" en Belgique', *Les Cahier du C.A.C.E.F.* 114: 3–12.

Gobin, C. (1986) 'L'Etat belge et la problématique de l'environnement', *Courrier Hebdomadaire du C.R.I.S.P.* 1109: 1–39.

Govaert, S. (1991) 'Le débat sur le Verzuiling en Flandre', *Courrier Hebdomadaire du C.R.I.S.P.* 1329: 69–79.

Hellemans, S. (1990) *Strijd om de moderniteit*, Leuven: Universitaire Pers K.U. Leuven.

Hooghe, M. and Rihoux, B. (2000), 'The green breakthrough in the Belgian general election of June 1999', *Environmental Politics*, 9, 3: 129–36.

Huyse, L. (1971) *Passiviteit, pacificacie en verzuiling in de Belgische politiek*, Antwerpen: Standaard Wetenschappelijke Uitgeverij.

— (1981) 'Political conflict in bicultural Belgium', in A. Lijphart (ed.), *Conflict and Coexistence in Belgium: the Dynamics of a Culturally Divided Society*, Berkeley, CA: Institute of International Studies, University of California.

— (1987) *De verzuiling voorbij*, Leuven: Kritak.

Kingdon, J. (1984) *Agendas, Alternatives and Public Policies*, Boston, MA: Little, Brown & Co.

Kitschelt, H. (1989) *The Logics of Party Formation. Ecological Politics in Belgium and West Germany*, Ithaca, NY, London: Cornell University Press.

Kitschelt, H. and Hellemans, S. (1990) *Beyond the European Left: Ideology and Political Action in the Belgian Ecology Parties*, Durham, NC: Duke University Press.

Lijphart, A. (1977) *Democracy in Plural Societies*, New Haven, CT: Yale University Press.

— (ed.) (1981a) *Conflict and Coexistence in Belgium: the Dynamics of a Culturally Divided Society*, Berkeley, CA: Institute of International Studies, University of California.

— (1981b) 'Introduction: the Belgian example of cultural coexistence in comparative perspective', in A. Lijphart (ed.), *Conflict and Coexistence in Belgium: the Dynamics of a Culturally Divided Society*, Berkeley, CA: Institute of International Studies, University of California.

Lijphart, A. and Crepaz, M. (1991) 'Corporatism and consensus democracy in eighteen countries. Conceptual and empirical linkages', *British Journal of Political Science* 21, April: 235–56.

Lorwin, V. R. (1974) 'Belgium: conflict and compromise', in K. McRae (ed.), *Consociational Democracy. Political Accomodation in Segmented Societies*, Toronto: McLellan & Stewart.

Michels, A. and Slomp, H. (1990) 'The role of government in collective bargaining: Scandinavia and the Low Countries', *Scandinavian Political Studies* 13, 1: 21–35.

Pijnenburg, B. (1984) 'Pillarized and consociational-democratic Belgium: the views of Huyse', *Acta Politica* 19, 1: 57–72.

Rihoux, B. (1994a) 'Les écotaxes – produits sur la scène politique belge (I)', *Courrier Hebdomadaire du C.R.I.S.P.* 1426: 1–36.

— (1994b) 'Les écotaxes – produits sur la scène politique belge (II)', *Courrier Hebdomadaire du C.R.I.S.P.* 1427–8: 1–29.

— (1994c) '"Ecotaxes" on the Belgian agenda, 1992–1994: a green bargain?', *Environmental Politics* 3, 3: 512–7.

— (1995a) 'Nieuwe sociale bewegingen in Franstalig België, eenheid in verscheidenheid', in S. Hellemans and M. Hooghe (eds) *Van Mei '68 tot Hand in Hand. Nieuwe sociale bewegingen in België, 1965–1995*, Leuven, Apeldoorn: Garant.

— (1995b) 'Ecolo en de nieuwe sociale bewegingen in Franstalig België', in S. Hellemans and M. Hooghe (eds) *Van Mei '68 tot Hand in Hand. Nieuwe sociale bewegingen in België, 1965–1995*, Leuven, Apeldoorn: Garant.

— (1995c) 'Belgium: greens in a divided society', in D. Richardson and C. Rootes (eds) *The Green Challenge. The Development of Green Parties in Europe*, London, New York: Routledge.

— (1997) 'Agalev, 1982–1997: entre marginalité et pouvoir', in P. Delwit and J. -M. De Waele (eds), *Les Parties Politiques en Belgique*, 2nd edition, Bruxelles: Editions de l'Université de Bruxelles.

Smits, J. (1979) 'De positie van de Belgische vakbonden in de politieke besluitvorming', *Politica* 29, 2: 155–77.

Steininger, R. (1977) 'Pilarization and political parties', *Sociologische Gids* 24, 2: 242–57.

Van den Brande, A. (1980) 'Neo-corporatisme en functioneel integrale macht', *Tijdschrift voor Sociale Wetenschappen* 25, 1: 33–60.

— (1987) 'Neo-corporatism and functional-integral power in Belgium', in I. Scholten (ed.), *Political Stability and Neo-corporatism*, London: Sage.

Van den Bulck, J. (1992) 'Pillars and politics: neo-corporatism and policy networks in Belgium', *West European Politics* 15, 2: 35–55.

van der Straaten, J. (1992) 'A sound European environmental policy: challenges, possibilities and barriers', *Environmental Politics* 1, 4: 65–83.

Vanderstraeten, A. (1986) 'Neo-corporatisme en het Belgische sociaal-economisch overlegssysteem', *Res Publica* 28, 6: 671–88.

Walgrave, S. (1994) *Nieuwe sociale bewegingen in Vlaanderen. Een sociologische verkenning van de milieubeweging, de derde wereldbeweging en de vredesbeweging*, Leuven: SOI/KU Leuven.

— (1995) *Tussen loyaliteit en selectiviteit. Over de ambivalente verhouding tussen nieuwe sociale bewegingen en Groene Partij in Vlaanderen*, Leuven, Apeldoorn: Garant.

Wilson, J. Q. (1980) 'The politics of regulation', in J. Q. Wilson (ed.), *The Politics of Regulation*, New York: Basic Books.

2   The role of the green movement in ecological modernisation

A British perspective

Peter Rawcliffe

The 10 years from 1985 to 1995 were a period of dramatic change in the environmental politics of Britain. This was marked both by a transition in the environmental policies of the main socio-economic and political institutions, and by a transformation in the political *gravitas* given to environmental issues within this polity. Galvanised by the growing pace and scientific understanding of environmental change, the green tide which swept through the country was driven by the perception that the environment had evolved into a legitimate, high profile and enduring political and policy issue. The environmental movement has been prominent in this process. This role reflects both the developing dynamics of environmental groups within the polity and, in turn, the broader changes in the environmental movement; particularly at the national level where environmental groups have become increasingly professional and well-resourced organisations with large mass memberships, management structures and sophisticated campaigning capabilities. How significant, however, have these developments been in terms of shifting policy onto a more sustainable footing? What, in essence, has been the role of the green movement in the ecological modernisation of policy in Britain?

This chapter seeks to explore this question, using a case study which assesses the development and impact of the campaigns of the national environmental movement in the development of road policy.[1] It begins by briefly outlining the main characteristics of the contemporary environmental movement in Britain, with particular attention given to the national groups which are involved in transport policy issues. The role of these national groups as movement actors is then developed and their interaction with the policy process explored. Using this theoretical perspective, the transport case study is introduced. Developments in road policy in those 10 years are reviewed and the structure of the policy process examined. Within this context, the changing nature and impact of transport campaigning is explored and an assessment made of the national groups' role in changing policy. The chapter concludes by briefly discussing the implications of this case study for the future development of the environmental movement.

The environmental movement in Britain

The British environmental movement has been generally credited as 'the oldest, strongest, best-organised and most widely supported environmental lobby in the world' (McCormick 1991: 34; Lowe and Flynn 1989: 268). These strengths are the product of the historical continuity, organisational diversity and leadership qualities which have characterised the development of environmental groups in this country. Importantly they also reflect the broader hold environmentalism has established on the public's consciousness and British society at large. This may be seen, for example, in the increase in the degree of sympathy expressed by non-environmental organisations and the wider public, the growth of green consumerism, the increased output of environmentally-oriented art and litera-ture, and even the sheer volume of news coverage on environmental issues – see, for example Porritt and Winner (1988). Clearly, an evaluation of the role of the environmental movement in the policy process must reflect this broader trans-formation and address 'why the particular groups of risks and dangers we call environmental have gained such purchase in our particular cultures and social circumstances' (Grove-White 1991a: 443). From this social-movement perspec-tive the real success of the environmental movement during those 10 years may be seen in facilitating, in articulating and in politically mobilising the broader green tide that swept through Britain. In turn, this emphasises the changing nature and activities of the national environmental groups through which this environmental pressure has largely been shaped in Britain.

The changing nature of the national environmental movement

The 1983–93 period was a remarkable period of change for the environmental movement.[2] In crude terms this is reflected in the growth in the movement's combined membership and resources. Between 1983 and 1988 membership rose from 1.8 million to over 4 million; nearly twice the size of the membership of the main political parties (Clarke 1990: 8). By 1990 it was estimated to be in the region of 4.5 million supporters – the 1993 level; the largest fifteen environmental groups had an income of £163 million and employed over 1000 full-time staff (Elkington 1990: 131). Since this peak, growth in membership and income has slowed and for some groups marginally declined. Despite this recent downturn, the membership and resource base of the environmental movement has, overall, increased significantly over the period. This has led to several important changes in the nature of the national movement.

Growth has been particularly marked in a relatively small number of large, membership-based groups. The groups include the Council for the Protection of Rural England (CPRE), Friends of the Earth (FoE), Greenpeace, the National Trust, the Ramblers Association, the Royal Society for Nature Conservation (RSNC), the Royal Society for the Protection of Birds (RSPB), the World Wild Fund for Nature (WWF) and the Woodland Trust. Together, these national groups account for the majority of members and supporters – with overlapping memberships between these groups a characteristic – as well as the main resource and organisational capacity within the movement. Information on the size, structure and activities of some of these key groups is provided in Table 2.1. This period has seen these groups in particular develop into more complex and increasingly corporate organisations characterised by administrative, marketing and fund-raising, media and legal sections. In several cases this has been a dramatic process with large increases in membership and resources allowing groups to grow exponentially in a period of 5 years from relatively small operations, employing fewer than twenty full-time staff, into organisations employing over a hundred staff. In turn, this has been marked by the increasing professionalisation of the national movement. Partly as a response to these changes, a more radical fringe has begun to emerge with North American direct-action groups such as Earth First!, Sea Shepherd, and the Rain Forest Action Group establishing active cells in Britain. Historically, this process of factionalism has been a common characteristic of the environmental movement providing a constant input of more radical groups, ideas and individuals.

As the national groups have grown and matured as organisations, the nature of their campaign and policy work has changed. In part this is the product of the changes in the political landscape with both government and industry more receptive to environmental issues per se. However, it also reflects the changing approaches of the groups, which has coincided with the resource increases and organisational developments described. This change can be seen in the shifting balance between conflict and more consensual styles, as campaigns have grown from simply aiming to raise the public consciousness of the issues to increasing the legitimacy of their case and on offering solutions rather than slogans. To this end, the increased resource base of the groups has enabled campaign and policy teams to be established, more sophisticated campaigns to be mounted and a greater emphasis to be placed on research. At the same time stronger relationships between groups have emerged. These are characterised by patterns of coordination and exist both formally through affiliated and umbrella groupings such as Transport 2000 and Wildlife Link, campaign alliances, issue and information exchange networks, and even the sharing of membership lists; and more informally through regular contacts and a 'seamless network' of individual friendships (Grove-White 1991b: 35). Despite the development at times of territorial disputes between some of the groups over campaign issues and competition for both members and resources, the development of these intergroup relationships has been important in the emergence of the environmental movement as an increasingly effective political force.

Towards a theoretical perspective

In Britain, the changing nature of the environmental movement may be usefully explored using elements of both the identity and resource mobilisation theories from the social movement literature.[3] From this perspective, environmental pressure may be concepualised as the product of an environmental movement

Table 2.1 Organisational profiles of selected British environmental groups (1993)

	Year started in Britain	Membership		Income (£)	Size of staff
		No.	% change (1985–93)		
RSPB	1889	850,000	+118	28,277,000	950
RSNC	1912	299,880	+39	2,852,000	56
CPRE	1926	46,000	+49	3,004,200	37
WWF	1961	230,000	+129	19,233,395	195
FoE	1970	230,000	+730	3,643,421	93
Transport 2000	1973	–		93,721[a]	3–4
Greenpeace Ltd	1977	411,000	+722	6,413,336	113
ALARM-UK	1990	–		–	3
Earth First!	1991	1,000[b]		–	–

Source: RSNC data.

Notes: [a]1992 data; [b]An estimate of the number of people/supporters ready to take non-violent action against road schemes. All data supplied by the groups.

RSPB [Royal Society for the Protection of Birds]: membership-based national bird conservation organisation which owns and manages 118 nature reserves and undertakes research, lobbying and campaigns. UK member of Birdlife International.

RSNC [Royal Society for Nature Conservation]: a partnership of forty-seven independent county trusts and fifty urban wildlife groups. These own and manage nature reserves. The national organisation provides coordinating services and collectively campaigns for wildlife protection.

CPRE [Council for the Protection of Rural England]: national organisation with forty-five democratically affiliated county branches and individual members. Undertakes research, lobbying and campaigns focusing on planning and countryside issues.

WWF [World Wildlife Fund for Nature]: membership-based national organisation which is part of WWF International. Important fund-raising role which has been complemented in recent years by increasing emphasis on research, lobbying and campaigns on a range of issues.

FoE [Friends of the Earth]: democratic membership-based organisation with 330 local groups and individual members. Part of FoE International. Combines media-oriented campaigning with research and lobbying skills across a range of environmental issues.

Transport 2000: broadly-based national federation of environmental and consumer groups, trade unions and local authorities which campaigns on national environmental and transport issues. Members include CPRE, FoE, RSNC, WWF and RSPB.

Greenpeace Ltd.: highly-centralised national organisation whose large supportership provides financial independence and backing for its direct actions, high profile research, and legal capacities on a number of environmental issues.

ALARM-UK: national alliance, support and information network of some 250 local community groups opposing road schemes. Staffed by volunteers.

Earth First!: anarchic grouping of individuals based in some forty-five cells which practice non-violent direct action, often of an illegal nature, against road schemes and other environmental threats. Derived from the North American Earth First! group.

responsive both to broader change in society and the dynamic nature of its environmental groups. In linking wider social processes with the socio-economic and political structures and by emphasising the processes or mechanisms by which environmental protest is socially constructed through both the formation and the activities of environmental groups, this framework therefore balances elements of both structure and agency (Niedhardt and Rucht 1992: 438; Jamison *et al.* 1990: 6). It is presented in Figure 2.1.

The national environmental groups, as the main organisational manifestation of the wider movement in Britain, have a key role in both mobilising the resources for protest and also in shaping the nature and form of this protest. As Figure 2.1 illustrates, this process of mobilisation can occur in two dimensions. First, the groups, in defining environmental problems and articulating an environmental agenda, assign meaning to and interpret events. When this process mirrors or gives symbolic form to the tensions within society which underpin its environmental aspirations and fears, mobilisation of potential supporters can occur (Grove-White 1991a: 441; Klandermans 1991). As Peter Melchett, Director of Greenpeace, observed:

> NGOs tap into the general level of (environmental) anxiety. People actually lose things and experience the effects of pollution or agricultural development or motorways [...] if the NGOs are doing their job well, they can do no more than reflect that reality.
>
> (Melchett in Grove-White 1991a: 446)

This consensus mobilisation can be manifested in the adoption of new lifestyles. More typically, however, it results in individuals joining or giving resources to established environmental groups in response to specific campaigns or recruitment drives. As well as mobilising resources, environmental groups also contribute to the definition of issues and agendas. Snow and Benford, in outlining this issue-framing process, argue, for example, that the strategy of such groups is consciously or intuitively planned in ways 'to mobilise potential adherents and constituents, garner by-stander support and demobilise antagonists' (Snow and Beford in Klandermans (1991: 33)).

Second, through their campaigns, environmental groups articulate goals and legitimise means of action and specific policies. Three types of campaign can be differentiated, although each may feature as part of a larger campaign. Policy and programme campaigns focus on specific policies (such as road building) or the practical outcomes of a policy (such as specific proposals for a road). They are generally directed at policy-makers – notably government and industry. In contrast, lifestyle campaigns are largely directed at the individual and are concerned with consciousness raising, personal responsibility, and empowerment. All three types of campaign can potentially be catalytic. In the short term, by raising awareness, they can increase the reservoir of potential recruits to the movement. In the longer term, in mobilising pressure and in raising public awareness, they can create the conditions for, and on occasion achieve, policy

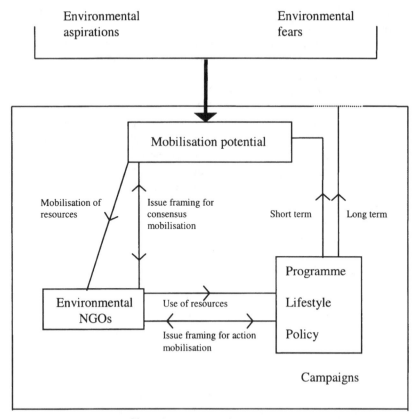

Figure 2.1   Towards a construction of environmental pressure.

change. This influence is revealed in the success of specific campaigns, legislative lobbying and political access, responses in institutional structures, and changes in the rhetoric and priorities of policy actors and interests.

Importantly, this period of mobilisation and organisational development has been mirrored and in part driven by the growing ability of the environmental movement, and particularly the national environmental groups, to develop and articulate these concerns within the polity of institutions, alignments and ideologies through which the nature and momentum of policy is shaped. Important components of this political opportunity structure include the political-administrative system and the executive bodies, the agents of control, in particular the courts; the political parties and interest groups; the mass media and the wider public (Niedhardt and Rucht 1991: 444). The groups' increasing ability to exploit this milieu reflects the changing spectrum of what Gamson and Meyer (1992: 4) usefully distinguish as stable and volatile elements.

The former aspects are fixed and deeply embedded in both the political institutions and wider society. For environmental groups, the impact of these elements of the opportunity structure has been extensive and is seen in both their form and strategy. Particularly important in the British context are policy communities.[4] These exist in many areas of Britain's segmented policy structure where a closed policy subsystem of stable actors share common values and goals, shape the rules of the game, influence membership and determine policy direction (Smith 1993; Richardson *et al.* 1992: 158). Reflecting this executive concentration of power, environmental groups have developed centralised, national structures based predominantly in London and in the case of Greenpeace and, for a period, Friends of the Earth fronted by establishment figures. Similarly, other groups – including WWF, RSPB and RSNC – receive royal patronage. Reflecting the increasing internationalisation of environmental policy, all these groups are also part of European or global federations, alliances or networks and, in the case of WWF and Greenpeace, part of a larger international organisation (Rucht 1993). In turn, national groups have pursued reformist strategies with groups incorporated into decision-making structures and gaining some degree of insider status. At times, this strategy has been complemented by media-based campaigning, including non-violent protests, publicity stunts and direct action – see, for example, Lowe and Goyder (1983: 62). As Rüdig *et al.* (1991: 139) observe, this ability of the British political and administrative system to accommodate and absorb new issues and groups has meant that 'movements effectively operate as pressure groups within established policy making systems, and therefore tend to fit the established analytical frameworks of pressure groups'.

The latter, volatile, components of the political opportunity structure are transient and shift with events, issue-attention cycles and the personal qualities of political actors; thus opening and closing opportunities for interaction in the policy process. It is these elements which influence rates of mobilisation and the interaction between movement strategy and these 'windows of opportunity': the 'opportunity for advocates of proposals to push their pet solutions, or to push attention to their special problems' (Kingdon 1984: 173). National environmental groups have largely developed to exploit or transform existing agenda opportunities and, on occasion, create new ones. Through public mobilisation and their campaign activities, environmental groups are engaged in two strategic processes (Kingdon 1984; Smith 1993). The first is the continual questioning of the dominant agenda. The second is the softening up of policy networks, experts and the public through redefining issues in new ways and by gradually establishing the environmental agenda as an institutionalised part of the policy process. This process of issue definition is a key influence on the stability and instability of policy communities; it and may lead to divisions and challenges both within and between these communities and to new actors emerging in more loosely fragmented issue networks (Baumgartner and Jones 1993; Smith 1993; Richardson *et al.* 1992). Important factors determining the ability of environmental groups to maximise agenda opportunities include both the resources and flexibility of the organisations and the skill and intuition of group leaders and individual campaigners. At the same time, the agenda opportunity structure influences the nature of environmental issues developed by the groups. For example, the observable reliance on the media arena by environmental groups has in part resulted in concentration on ambulance-chasing environmentalism. Echoing media structures, this has led to the mediagenic simplification of complex issues and emphasised their conflictual and competitive aspects. In turn, groups have tended to narrowly focus on the physical environment rather than the interlinked issues of justice, equity and welfare.

A case study: inroads into roads policy?

In Britain, transport policy and specifically the debate over road building and car use has rapidly emerged since 1988 to become an established part of the broader environmental agenda. Underlying this change has been the increasing acceptance that balancing the environmental costs, particularly of the creeping dependence on road-based transport, with the needs of the modern economy, represents one of the key challenges in moving towards sustainable growth both in this country and internationally. The government's own Environment White Paper *This Common Inheritance*, for example, identified the resolution of the conflict between mobility and what is best for our towns and countryside as one of the three 'critical issues' in the development of a sustainable future (HM Government, 1990: par. 2.6).

Despite the emergence of the environmental agenda, policy remains essentially committed to demand-led road building and the continuation of the great car economy on which the high-mobility society is based and to which much of the environmental critique is directed. As David Hall notes: 'the transport policies pursued since the appearance of the White Paper, have, in the main, borne little relationship to its stated aims' (Hall 1993: 8). At the same time, policy is increasingly and often insidiously being justified on environmental grounds. For example, in 1993 the Secretary of State for Transport proudly justified the expansion of the roads programme under his leadership suggesting 'in fact I am a very strong environmentalist and I would definitely defend what I'm doing in that respect'.[5] The outlined theoretical framework can be used to explore the dichotomy between the emergence and expression of the environmental agenda in the institutional processes and practice of transport policy in Britain. To this end the rest of the chapter examines the key policy structures, considers the main developments in the opportunity structure in this period, and assesses the changing role and influence of national environmental groups in the development of transport policy.

Transport policy and the development of the great car economy paradigm

In Britain, transport policies pursued since the seminal Buchanan report *Traffic in Towns* (DoT 1963) have had two main aims: to build enough road capacity to meet forecast demand and to reduce public spending on public transport. Despite protests against road building in the 1970s and the presence of a small but vocal group of maverick experts, the rise of car ownership and traffic growth has led to a transport policy becoming synonymous with roads policy.[6] The dominance of the great car economy in the political opportunity structure is well documented, notably by proponents of alternative policy paradigms.[7] This reflects the cumulative result of five interrelated factors.

First, there is the historical development of state intervention in transport. This has resulted in road building becoming legitimised as a key government activity and public transport and planning being increasingly marginalised (Potter 1982). After 1979 the right-wing ideology of the successive Conservative governments – seen in its policies of deregulation, privatisation and the promotion of the enterprise culture – reinforced this trend (O'Riordan 1988). Secondly, there are the departmental policy structures. These – reflecting the British style of policymaking (Jordan and Richardson 1979: 44) – are based on a stable, compartmentalised and producerist policy community centred on the Department of Transport (DoT).[8] Reynolds, for example, observes 'the extent to which within the DoT, ideology, procedures, personnel and resources have become focused on the management and construction of roads' (Reynolds in Kay 1992: 5). Next, there is the power of the road lobby. This is a classic example of a highly effective and powerful insider group. Essentially corporatist in nature, this client grouping consists of a wide range of traditional economic interests[9] which, in conjunction with the professional engineering bodies, have a substantial and long standing share in formulating and administering government policy (Hamer 1987; Dudley 1983: 104). Fourthly, there is the importance of the road industry and growth in traffic for the British economy. It is this strong relationship, together with the increasing dependency of the state on road-generated revenue,[10] that has been a major factor in the DoT's roads programme consistently remaining a Cabinet and Treasury favourite in terms of public-spending rounds (Banister 1992: 205). Finally, the idea of the motorway society has become influential. The pursuit of policies aimed at increasing car ownership and use had their own momentum. This is seen both in society's social and structural accommodation of motorised traffic – as in the remaking of our mobility and land-use patterns and in the growing psychological dependence on the car (Stokes and Hallett, 1992; Roberts 1990).

Through this opportunity structure, the policy-influenced and directed momentum of roads policy may be seen as the product of systems of decision-making and non-decision-making, as well as the single decisions or what Lukes calls overt or first-dimension power (Lukes 1974). This in turn highlights aspects of second- and third-dimension power which emphasises, respectively, 'the control of political access and agendas, whereby certain groups are excluded from decision making and certain issues and policy options from consideration; and ideological control whereby certain interests in society enjoy an overriding legitimacy' (Lowe and Rüdig 1986: 529). This mobilisation of bias, which has influenced the direction of transport policy, has become both self-perpetuating and reinforcing. The implication of this is most clearly seen in stark political terms. Changing the emphasis of policy towards any form of restriction in car use, for example through road pricing, continues largely to be seen as both economic and electoral suicide. This attitude is exemplified by the Secretary of State for Transport's familiar justification for the allocation of road schemes in 1992: 'There will be a substantial road building programme because we see more and more people wishing to move around the country and we hope to cope with that' (House of Commons, 15 December 1992). Despite this prognosis, the 1988–95 period has seen the dominant paradigm being increasingly challenged, as transport has developed into a major public issue, and the broader greening of the 1980s has seen environmental concerns increasingly reflected in institutional practice and policy.

Going green? The changing opportunity structure

By the late 1980s unprecedented traffic growth and levels of congestion bore witness to an increasing crisis in transport policy, which directly threatened the dominant road-building consensus within the policy community (Ward *et al.* 1990: 232). A sustained period of economic growth and rising car ownership, combined with a government strongly committed to the car and ideologically opposed to both integrated land-use planning and expenditure on public transport, was too much for an ailing transport system graphically described, for example, by Rose (1990: 185):

> Easter 1989 saw the M4 clogged up from London to Bristol in a 125-mile traffic jam. The Severn Bridge, pounded by the unplanned-for juggernaut traffic, was under almost constant repair. Large chunks of the M5 and M6 began too fall to pieces, as did the M1.

The wider mood was similarly pessimistic. The 1990 Conservative Party Conference, for example, had seen ninety-four motions debated on transport – more than any other subject bar law and order. Similarly, the conservative *Daily Telegraph* (5 September 1991), commenting on plans to upgrade the M25 to fourteen lanes, dismissed the scheme as a pointless waste of money that 'will ultimately have little effect on congestion' (Joseph 1990a: 326). After 1990, fears for the future of rail services in the light of the long-delayed plans for rail privatisation also added a further dimension to deepening public anxiety over the direction of transport policy.

In 1989, the *Roads to Prosperity* White Paper announced a 'greatly expanded' £12 billion roads programme designed to meet the staggering National Road Traffic Forecast's predicted 83–142 per cent rise in traffic by the year 2025 (DoT 1989a,b). As Goodwin et al. (1992: 2) observe, the release of these papers began to transform the debate on transport policy as 'month by month local transport planners were working out the consequences for their own area and professional institutes were considering the consequences for their own discipline and role'. Against a background of rising public concern about green issues, the policy ramifications of the *Roads to Prosperity* White Paper not only challenged conventional transport thinking but also ensured that transport would remain high on the political agenda. Chris Patten, then Secretary of State for the Environment and in charge of producing Britain's first White Paper on environmental policy, reflected public anxiety when he talked, for example, of the need for fresh ideas in the light of the likely environmental impacts of the 'unacceptably high' traffic forecasts (*Guardian*, 28 September 1990). From this period onwards, concern over the direction of transport policy began to be reflected both in policy and within the policy community. Two sets of changes can be seen: the increasingly environmental context for transport policy, often marked by the involvement of institutions and groups not previously associated with policy; and the development of new thinking within parts of the existing policy community.

The new environmental context for transport policy

The beginnings of the integration of environmental and transport policy has provided an important locus for the environmental agenda within the transport community. This has been driven by the government's annual Environmental White Paper review process and, since the Earth Summit in June 1992, by both the Agenda 21 process stemming from UNCED and the more specific commitments to carbon-dioxide reductions throughout the Framework Convention on Climate Change. This directly led to conflict in policy-making marked by the growing tension between the Department of the Environment (DoE), responsible for Britain's environmental policy (the White Paper Unit within the Department coordinated the Environment White Paper review process), and the DoT over the direction of transport policy.[11] Similar tensions emerged between the European Union and the British government throughout the 1985–95 period over exhaust emissions and the environmental impact appraisal of road schemes.[12]

For the DoT, which prided itself as being the largest planter of broad-leaved trees in the country and strongly argued that the roads programme is good for the environment, this period was increasingly difficult. While the policy impact of the departmental green minister, appointed as a result of the White Paper process, was minimal, the establishment of a Transport Policy Unit to consider policy matters may be seen as of greater long-term importance. Through the reorientation towards internal debate that this unit generated, certain internal groupings, mainly within a new guard, were becoming increasingly estranged from the Department view. As Kay (1992: 65) notes 'even in Marsham Street doubts were emerging: the Department has become a dinosaur, under attack from ever more unlikely directions'.

The emergence of the new realism

The policy community has also begun to respond to the greening of the policy agenda, with the new emphasis clearly seen in the language of the debate. While change has been driven by the need for green credentials, the process is also in part the product of the development of new thinking within parts of the policy community which Goodwin *et al.* (1992) have called the 'new realism'.[13] Changes in personnel and the establishment of working groups, policy units and even independent research foundations have enabled the policy community to participate more fully in the developing public debate through their public statements and information services. Importantly, these developments have ensured that the community can contribute effectively at several levels of the policy process including the annual Environment White Paper review, the ongoing work of the Climate Change Committee and various EC Green Papers such as *Sustainable Mobility* in 1992. During this period the policy community also become more open generally to environmental arguments and environmental actors. This change is seen, for example, in the informal sharing of information and reports between these organisations and environmental groups and their greater acceptance of the legitimacy of such groups to take part in the policy process.

The development of environmental campaigning

The transport campaigns of environmental pressure groups have developed in response to this changing opportunity structure, particularly post-1989 when, for the reasons outlined, the issue surged back into the public spotlight. In part this reflects the rapid maturing of these groups as political actors and their development into professional organisations. Transport has also developed into a high-profile campaign issue for the environmental movement, as manifested by the development of a broad coalition of interests. Important in these developments has been the process of consensus-building structures such as the Transport Activists Roundtable. Established by Transport 2000 primarily as an information and support network, this forum has grown from a nucleus of nine national environmental and transport groups in early 1990 to over twenty groups, ranging from Earth First! to the Countryside Commission, by 1992. As, for example, Stephen Joseph, Director of Transport 2000, observed: 'there is a fairly coherent view about what is wrong with current road/transport planning (*partly as a result of the groups working together within the Roundtable*) and the outline of what to replace it' (Notes from Transport Activists Roundtable meeting, 26 November 1990; emphasis added). This consensus may also be seen in the *Roads to Ruin* Statements of 1989 and 1992, which both challenged the 1989 *Roads to Prosperity* White Paper by presenting 'collectively and coherently [...] the arguments the government has failed to address' (CPRE 1989: 5).

Importantly, groups with no previous track record in transport campaigning – such as Greenpeace, WWF-UK and the conservation organisations – have also for the first time begun to tackle the issue. This has increased the overall level of resources at the national level and has complemented the particular strengths of the groups – such as FoE, CPRE and Transport 2000 – traditionally associated with transport campaigning. Even for groups already active on transport, surging membership and income has meant more resources for campaigns. Thus from the 'three and a half people meeting in a London pub celebrating occasional victories' at the beginning of the 1980s, by 1992 there were about twenty full-time staff at the national level dealing specifically with the transport issue. The presence of these extra campaigners, plus the researchers and other support staff deployed, has allowed a much higher campaign profile to be established.

At the national level campaigning on roads, while spanning a variety of issues from the promotion of road safety to opposition to heavier lorries and bypasses, has traditionally focused on campaigns against specific road proposals, as in the mid-1980s' protests against the Okehampton and Lynhampton bypasses and the M40 Oxford to Birmingham link. Given the changing opportunity structure, the emphasis since 1989 has however increasingly been policy oriented. As Stephen Joseph, Director of Transport 2000, observed:

> [G]roups have tended in the past to oppose specific routes for specific new roads, rather than the roads themselves. The new roads programme and the impact in terms of pollution and the demand for land of an increasing vehicle population seems likely to change this. It will lead those in conservation and other environmental groups to challenge the direction and bias of overall transport policies and to propose alternative, more environmentally sustainable policies [...].
>
> (Joseph 1990b: 20)

In part this change reflects the development of a campaigning infrastructure that has enabled more intensive policy-level campaigns and research to be undertaken. FoE's *Roads Campaigner* in 1992, for example, observed that the increasing sophistication of support services at FoE – ranging from campaign teams, researchers, administrators, communications and media support – meant that there was more time available to concentrate on the strategic nature of the campaign. Symptomatic, however, of the increasing accessibility of the policy process and the transport community is the increased time now devoted to meetings with ministers and their officials, giving evidence to select committees, responding to various policy statements and White Papers. Groups have also developed their relations with the media and are increasingly adept at planting stories in elements of the press and in setting wider media agendas. In many respects this has been a learning process for the national groups, especially for those which were new to transport campaigning. Notable as an octopus issue, transport cuts across a range of issues and has been structurally difficult for the groups to get right, often resulting in the fragmentation of elements of the transport campaign across other campaigns. Within these structural constraints, integration and the formation of effective campaigns has at times been problematic, while the obstacles encountered by some of the organisations when working together have, at times, worked against partnerships at both the national and local level.

At the local level, Nimby (not in my back yard) protests against road building have traditionally been a strong element of campaigns. As the massive roads programme outlined in the 1989 *Roads to Prosperity* White Paper has gradually been implemented, these have grown in importance. In part this change reflects the involvement and resources of nationally recognised and respectable environmental groups which has empowered both a greater number and a wider range of people, enabling them to get involved. However, limited resources have meant that national campaigns have had to be targeted on a few selected schemes such as the M25 widening.

Of more widespread importance since 1989 in strengthening grass-roots protest has been the development of considerable expertise and support at the local level. Two key elements in this change may be described. First, Alarm-UK,[14] building on its success in fighting against road proposals in London, has established a national anti-road-building network which acts to support and encourage new groups. This network has been particularly effective, both in empowering local groups to take action and in bringing together these local groups at the regional level where they are fighting elements of what amounts to the same road scheme. Second, there has been the emergence of more radical direct-action groupings such as the Dongas Tribe,[15] Earth First! and Road Alert. Since the Twyford Down protests in 1991[16] and the emergence of a young, mobile and radical generation of protestors, non-violent direct action has become a third force in campaigning. Its emergence has had a direct impact on the roads programme. This was seen particularly in the case of Oxleas Wood in London, where the threat of mass protest was seen as critical in the government's abandonment of the proposed East London River Crossing road improvements. Together, these developments at the local level have been important both in increasing the political and financial costs involved in road construction and in adding to the national pressure for change in policy direction. As John Stewart of Alarm-UK observed of the national groups: 'For all the talking they do to politicians, DoT officials and journalists – unless there is a grass-roots movement out there asking for the same sorts of things – they're not going to be listened to' (personal communication, 6 December 1993).

The impact of environmental pressure: the changing policy climate

In many respects it is too early to judge how effective environmental pressure has been in shifting transport policy onto a more sustainable footing. Despite changes in the opportunity structure within which the environmental groups were operating, the increasing institutionalisation of the environmental agenda in the transport debate and the greater effectiveness of the environmental groups in participating in the decision-making process, there had yet to be, by the mid 1990s, a policy spiral in transport policy away from the dominant road-building paradigm. Indeed, the greening of transport policy has been more cosmetic than substantial, resulting more in a change of language and a few glossy documents.[17] Perhaps most significantly the £12 billion national roads programme announced in *Roads to Prosperity* (DoT 1989a) was progressively expanded reaching a nominal £23 billion by the beginning of 1994.

There were however the first signs of change. The publication in January 1994 of the government's *Sustainable Development Strategy* brought confirmation that the government, in now arguing that continued traffic growth 'would have unacceptable consequences for both the environment and the economy of certain parts of the country, and could be very difficult to reconcile with sustainable development goals', had perceptively shifted its perspective (HM Government, 1994: par. 26.17). In turn, successive recent Secretaries of State for Transport called for a proper public debate on the issues and promised a future White Paper on transport policy.[18] While these developments did not produce any shift in the direction of policy away from the great car economy paradigm, the gradual change in the language of policy is important – and looks likely to lead to the greater integration of land-use and transport planning, particularly in respect to out-of-town retailing and a scaling-down of parts of the current road programme (although significantly not a redirection of transport investment towards public transport). In line with the current trend towards indirect taxation there are also likely to be further increases in fuel prices as part of the government's policy of reducing carbon dioxide emissions and the implementation of various forms of road pricing. The 1994 reports of the influential Royal Commission on Environmental Pollution (1994) and the Standing Advisory Committee on Trunk Road Assessment (1994) are likely to strengthen the case for these measures.[19]

These shifts were symptomatic of the flux within the decision-making community over the direction of transport policy during the Major premiership. This instability may be seen in both the growing tensions between the DoE and DoT over the direction of policy and the increasing pressure on the DoT from the Treasury to reduce expenditure. The inclusion of peripheral groups such as the powerful medical lobby – increasingly concerned about the health impacts of cars and car use – in the policy process is also notable. At the same time the roads programme began to cause deep resentment in the Tory heartlands of middle England. As a result doubts emerged within both the pro-Conservative national newspapers, such as the *Daily Telegraph*, and the powerful grass roots of the Conservative Party concerning the direction of transport policy. Visible signs of this tension were seen in the increasing number of Conservative MPs who were beginning to openly challenge government transport policy. They included those whose constituencies were affected by the M25 widening programme and other road schemes in the Home Counties and by threats to rail services caused by privatisation.

As both an increasingly legitimate part of the debate and an important source of scientific research and new thinking, the national environmental groups have successfully exploited the flux inherent within the opportunity structure to both push and shape the nature of policy change. As well as pushing issues such as habitat and landscape loss, global warming and the impact of air pollution on health onto the transport agenda, the national groups have also on occasion been able to produce identifiable changes in policy on certain winnable issues. Significantly more influence was achieved with the DoE – as seen, for example, in the production in 1994 of *Planning Policy Guidance Note 13: Transport* (DoE 1994), which provides national guidance on integrating land-use and transport planning – where access for the environmental groups has traditionally been greater, than the DoT.[20]

At the programme level, victories have been won, especially where schemes have affected significant numbers of people – such as in London where the government in 1990 backed down from the proposals produced by the London Assessment Studies – and large areas of the country – as in the case of the proposed East Coast Motorway. Clearly, the ongoing reorientation of the road-building programme away from urban areas towards bypasses and the widening of existing trunk roads and motorways is a tactical recognition of the successful mobilisation of people on the ground. Even the planned road-building programme is under increasing threat, as reported in a leaked DoE document, which detailed the serious opposition to some 243 controversial road developments. The increasing success of protests of this nature, as in the case of the A3 at the Devil's Punchbowl, the Hereford Bypass and M62 link road has given the environmental movement its most powerful campaign victories. Even failures, most notably at Twyford Down, have been highly symbolic and have been catalytic in terms of mobilisation and in creating a groundswell of feeling against road building. However, despite these undoubted successes and its revision in the November 1994 Budget from £23 to £18 billion, the roads programme was still the largest in the DoT's history. From this perspective, the policy changes that took place remain relatively small and occurred within the dominant paradigm.

Conclusions: prospects for ecological modernisation

Given the growing dependence of modern societies on transport and the increasing environmental costs this incurs, the role and influence of environmental groups in influencing the development of transport policy in Britain has provided a useful case study with which to explore the part played by the green movement in the ecological modernisation of policy. This is an undoubtedly important role. As part of the surge in environmental concern, environmental groups have been effective both at framing the anxieties over congestion, pollution and road building in Britain as a significant element of a broader environmental agenda and in mobilising protest, particularly against road building. At the national level, this impact is seen in the increasing success of groups in raising awareness and support through their transport work. In 1993 the CPRE, for example, increased its membership and raised over £100,000 from their transport campaign appeal, while FoE's Road to Ruin Appeal has been similarly successful. This success has been mirrored at the local level by the development of a broadly-based coalition of both respectable and more radical anti-road groups and individuals. Through these processes the environmental movement has undoubtedly been important in augmenting and shaping the climate in which transport policy has been debated during the 1990s. This climate has in turn led to the observable, if unquantified, flux within the policy community and in policy.

Together, these developments suggest that in the short term elements of the roads programme may be stopped while other aspects of transport policy may be made more environmentally benign. Little in the short to medium term, however, suggests that the wider transport problem embedded within the dominant great car economy paradigm may be solved. This assessment is based on an acknowledgement of the current limits to ecological modernisation seen in terms of the prevailing structures of power within society through which environmental pressure is both expressed and constrained. For the transport conundrum this power is revealed in the embedded nature and socialisation of car ideology within the current political process, the producerist policy community, the economy and wider society.

Adapting Sabatier's model of policy change, the manifestation of power within this political opportunity structure may usefully be described in terms of three overlapping frameworks or belief systems (Sabatier 1988). The first framework is one of ideology and deep core beliefs. This includes the adherence to the economic-growth paradigm and the individual's right to travel. The second framework is one of basic political values or strategies. This includes, for example, the political philosophies inherent in the great car economy and the promotion of car ownership. The third framework is one of specific policy measures, such as the scale of road building, tax structures and road pricing. It is in this framework that the new realism in transport policy may be clearly observed. Extending Luke's analysis (Lukes, 1974), resistance to the process of ecological modernisation is greatest at the level of deep core beliefs where third-dimension power is particularly manifested and the political opportunity structure is most stable. Resistance to this process is least at the level of specific policy components where first-dimension power is manifest and pluralist processes are most visible. It is at this level that the political opportunity structure is most volatile and where the environmental movement has been most effective in challenging specific elements of policy and in raising public and media awareness.

Through this opportunity structure, the nature of environmental pressure has, as noted, also been influenced by the process that Herman (1992: 890) has described as 'the "rate of exchange" between a movement's ideological rigidity and its actual political efficacy'. As an increasingly effective movement for influencing policy within the British polity, the national environmental groups have increasingly adopted norms of organisational behaviour and rational approaches to transport campaigning which, for the most part, do not challenge the deeper core frameworks inherent in the political opportunity structure. As a result it can be argued that in their success in mobilisation and influencing policy, the national groups have paradoxically become a less radical force for ecological modernisation. Perhaps symptomatic of these constraints is the absence for the most part of lifestyle campaigns which question some of the deeper core beliefs and freedoms associated with the great car economy. Given their dependency on their membership, this is perhaps taking the environmental groups into uncertain territory. As a driver survey in 1992 commented:

Evidence to date, however, suggests that consumers will not swallow a bitter environmental pill. They will not stop consuming but will express their concern by buying more environmentally friendly products when they are easy to buy, to use, do not have significantly worse performance and, importantly, do not have a significant price penalty. The relatively slow uptake of unleaded petrol in the UK, despite significant cost benefits, supports this view.

(Lex Report on Motoring 1992: 136)

Given the continuing lack of real change in transport policy and the increasing dependency of modern society on the motor car, the national environmental groups may now have to rethink both their structures and strategies if they are going to become part of the catalyst towards the social transformation and ecological modernisation that is required.

Notes

1   This chapter draws from P. Rawcliffe (1998), *Environmental Pressure Groups in Transition*, Manchester: Manchester University Press; and is based on postgraduate research undertaken by the author between 1991 and 1994 at the School of Environmental Sciences, University of East Anglia. It was funded by the Economic and Social Research Council.
2   Commentaries on different aspects of this period are provided by McCormick (1991); Rootes (1992); Rawcliffe (1994).
3   The development of this approach to social movements has been argued for by, for example, Klandermans (1991) and Niedhardt and Rucht (1991).
4   This is a type of what Rhodes and Marsh (1992) term a policy network. These authors also distinguish issue networks which, in contrast to policy communities, are characterised by large numbers of participants with no restrictions on entry.
5   This was part of the Secretary of State for Transport's characteristic reply to a question concerning pollution from motor vehicles on BBC Radio Four's *Question Time*, 23 February 1993.
6   This emphasis is indicated by the production of White Papers, the official statements of government policy. The last transport policy White Paper was published in 1977 (DoT 1977). In contrast, roads White Papers were released in 1983 (DoT 1983), 1985, 1987 and 1989 (DoT 1989a).
7   For a review of the development of and current nature of transport policy see, for example, Roberts *et al.* (1992); Kay (1992); Hamer (1987); Potter (1982); Adams (1981).
8   The DoT determines policy within the broad policy agenda of the government; sets the financial allocation for investment programmes (within a broad figure set by the Treasury with reference to the Department of the Environment); and manages the national trunk road network and its rolling infrastructure programme. In 1992, nearly 60 per cent of staff worked on the planning and building of roads and managing road traffic, compared with 1.5 per cent who worked on public transport (Joseph *et al.* 1992: 89; Kay 1992).
9   This grouping includes the British Roads Federation, the motor industry (represented by the Society for Motor Manufacturers and Traders), the bus operators, road haulage firms, the motoring organisations (including the Royal Automobile

Club and the Automobile Association), the road-construction industry and the oil industry (Hamer 1987: 9).

10   Road-use expenditure in 1989 contributed some £17 billion or 12 per cent of total exchequer revenues. This represents an increase of 62 per cent in real terms since 1979 (Banister 1992: 205). Increases in tax on fuel in the 1993 Budget as part of the government's policy of reducing carbon-dioxide emissions will yield a further £1.5 billion annually, reinforcing this trend in public-revenue generation (DoE (1993: 2).

11   Ironically, given their clear lack of communication and coordination, both these goverment departments share the same building in Marsham Street.

12   Reviews are provided by Rose (1990); Alder (1993).

13   This change, essentially a response to the predicted traffic growth, is based on the recognition that there is no physical possibility of increasing road supply at a level which approaches the forecast increases in traffic. The new consensus is therefore about 'a roads policy which focuses on real need, traffic management procedures, public transport systems, traffic calming and pricing mechanisms' (TEST 1991: 25). While embryonic, this consensus is a broad one encompassing many of the groups traditionally associated with the road lobby. As well as the motoring organisations these include local government associations and the various bodies representing the civil engineering, chartered surveyors, architectural and transport professions.

14   This grouping originally developed from a network of local groups as ALARM (All Londoners Against Road-Building Menace) in response to the London Assessment Studies.

15   This grouping is named after a network of ten-foot deep trackways found on Twyford Down. These, ironically perhaps, form part of one of the UK's oldest road systems.

16   Twyford Down, nominally one of the most protected landscapes in the UK, was in part removed in 1991–3 to make way for the M3 extension. This was after a 20-year struggle by local residents, the national environmental movement and, more recently, radical grass-roots groups such as the Dongas Tribe, as well as intervention by the EC.

17   Illustrative of this limited response, a written answer by the Secretary of State for Transport to a parliamentary question listing the environmental achievements of his Department in 1992–3 included the publication of *The Good Roads Guide*, aimed at professionals involved in the design of trunk roads to make them as 'environmentally satisfactory as possible'; to be followed by an updated *Manual of Environmental Appraisal, The Wildflower Handbook* and *The Design Guide for Environmental Barriers* (House of Commons, 15 April 1993).

18   This public debate took the form of six public speeches by the Secretary of State which were used to ask a series of questions on the major transport issues.

19   The Royal Commission on Environmental Pollution (RCEP) issued its report in October 1994 after nearly 5 years of investigation. The report defines objectives for a sustainable transport policy and includes more than a hundred recommendations. The report by the Standing Advisory Committee on Trunk Road Assessment, which highlights further flaws in the cost–benefit analysis procedures with which new road building has been officially justified, is likely to be equally significant in policy terms.

20   Two notable exceptions to this were the DoT's new policy on traffic calming and its increasing promotion of cycling.

References

Alder, J. (1993) 'Environmental impact assessment – the inadequacies of English law', *Journal of Environmental Law* 5, 2: 203–20

Adams, J. (1981) *Transport Planning: Vision and Practice*, London: Routledge.

Banister, D. (1992) 'Transport', in P. Cloke (ed.), *Policy and Change in Thatcher's Britain*, Oxford: Pergamon Press.

Baumgartner, F. and Jones, D. (1993) *Agenda and Instability in American Politics*, Chicago, IL: Chicago University Press.

Clarke, C. (1990) 'Improving the environmental performance of business: the agenda for business and the role of environmental organisations', unpublished M.Sc. thesis, Centre for Environmental Technology, Imperial College of Science, Technology and Medicine.

CPRE (1989) *Annual Report*, London: Council for the Protection of Rural England.

DoE (Department of the Environment) (1993) *Climate Change. Our National Programme for CO₂ Emissions*, addendum to the discussion document, London: HMSO.

—— (1994) *Planning Policy Guidance Note 13: Transport*, London: DoE.

DoT (Department of Transport) (1963) *Traffic in Towns: a Study of the Long Term Problems of Traffic in Urban Areas*, London: HMSO.

—— (1977) *Transport Policy*, Cmd 6836, London: HMSO.

—— (1983) *Policy for Roads*, Cmd 9059, London: HMSO.

—— (1989a) *Roads to Prosperity*, Cmd 693, London: HMSO.

—— (1989b) *National Road Traffic Forecasts (Great Britain)*, London: HMSO.

Dudley, G. (1983) 'The road lobby: a declining force?', in D. Marsh (ed.), *Pressure Politics*, London: Dent.

Elkington, J. (1990) *A Year in the Greenhouse*, London: Victor Gollancz.

Gamson, A. and Meyer, D. (1992) 'Framing political opportunity', paper to the American Sociological Association Annual Meeting, Pittsburgh, PA, August.

Goodwin, P., Hallett, S., Kenny, F. and Stokes, G. (1992) *Transport: the New Realism*, Oxford: Oxford University.

Grove-White, R. (1991a) *The UK's Environmental Movement and UK Political Culture*, Lancaster: Centre for the Study of Environmental Change, Lancaster University.

—— (1991b) 'The emerging shape of environmental conflicts in the 1990s', *RSA Journal* 129, 5419: 35.

Hall, D. (1993) 'Getting around – transport and sustainability', *Town and Country Planning* 62, 1: 8–12.

Hamer, M. (1987) *Wheels within Wheels – a Study of the Road Lobby*, London: Routledge & Kegan Paul.

Herman, T. (1992) 'Contemporary peace movements: between the hammer of political realism and the anvil of pacifism', *The Western Political Quarterly* 45, 4: 869–94.

HM Government (1990) *This Common Inheritance: Britain's Environmental Strategy*, Cm 1200, London: HMSO.

—— (1994) *Sustainable Development: the UK Strategy*, Cmd 2426, London: HMSO.

Jamison, A., Eyerman, R. and Cramer, J. (1990) *The Making of the New Environmental Consciousness: a Comparative Study of the Environmental Movements in Sweden, Denmark and the Netherlands*, Edinburgh: Edinburgh University Press.

Jordan, A. and Richardson, J. (1979) *Governing under Pressure – the Policy Process in a Post-Democracy*, Oxford: Blackwell.

Joseph, S. (1990a) 'Brand-new secretary, worn-out paradigm', *Town and County Plan-ning* 59, 12: 326.

— (1990b) 'Roads to where?', *ECOS* 11, 2: 17–20.

Joseph, S., Lester, N. and Hamer, M. (1992) 'The machinery of government', in J. Roberts, J. Clearly, K. Hamilton and J. Hanna (eds.), *Travel Sickness – the Need for a Sustainable Transport Policy for Britain*, London: Lawrence & Wishart.

Kay, P. (1992) *Where Motor Car is Master – How the Department of Transport Became Bewitched by Roads*, London: CPRE.

Kingdon, J. (1984) *Agendas, Alternatives and Public Policies*, Boston, MA: Little, Brown.

Klandermans, B. (1991) 'New social movements and resource mobilisation: the European and American approach revisited', in D. Rucht (ed.), *Research on Social Movements: the State of the Art in Europe and the USA*, London: Westview Press.

Lex Report on Motoring (1992), London: Lex Motor Group.

Lowe, P. and Flynn, A. (1989) 'Environmental politics and policy', in J. Mohan (ed.), *The Political Geography of Contemporary Britain*, London: Macmillan.

Lowe, P. and Goyder, J. (1983) *Environmental Groups in Politics*, London: George Allen & Unwin.

Lowe, P. and Rüdig, W. (1986) 'Political ecology and the social sciences – the state of the art', *British Journal of Political Science* 16, 4: 513–30.

Lukes, S. (1974) *Power: a Radical View*, London, Basingstoke: Macmillan.

McCormick, M. (1991) *British Politics and the Environment*, London: Earthscan.

Niedhardt, J. and Rucht, D. (1991) 'The analysis of social movements: the state of the art and some perspectives for further research', in D. Rucht (ed.), *Research on Social Movements: the State of the Art in Europe and the USA*, London: Westview Press.

— (1992) 'Towards a "movement" society?: on the possibilities of institutionalising social movements', paper presented to the First European Conference on Social Movements, Berlin, October.

O'Riordan, T. (1988) 'The politics of environmental regulation in Great Britain', *Environment* 30, 8: 5.

Porritt, J. and Winner, D. (1988) *The Coming of the Greens*, London: Fontana.

Potter, S. (1982) *Understanding Transport Policy*, Open University course on transport policy, section 3, Milton Keynes: Open University Press.

Rawcliffe, P. (1994) 'Swimming with the tide: the changing nature of national envir-onmental pressure groups in the 1990s', unpublished Ph.D. thesis, University of East Anglia.

Richardson, J., Malony, W. and Rüdig, W. (1992) 'The dynamics of policy change: lobbying and water privatisation', *Public Administration* 70, 2: 157–75.

Rhodes, R. A. W. and Marsh, D. (1992) 'New directions in the study of policy networks', *European Journal of Political Research* 21, 1–3: 181–205.

Roberts, J. (1990) *User Friendly Cities*, London: TEST.

Roberts, J., Cleary, J., Hamilton, K. and Hanna, J (eds.) (1992) *Travel Sickness – the Need for A Sustainable Transport Policy for Britain*, London: Lawrence & Wishart.

Rootes, C. (1992) 'Environmentalism: movement, politics and parties', *Environmental Politics*, 1, 3: 465–9.

Rose, C. (1990) *The Dirty Man of Europe – the Great British Pollution Scandal*, London: Simon & Schuster.

Royal Commission on Environmental Pollution (1994) *Transport and the Environment*, Cmd 2674, London: HMSO.

Rucht, D. (1993) '"Think globally, act locally?" Needs, forms and problems of cross-national co-operation among environmental groups', in D. Liefferink, P. Lowe and A. Mol (eds.), *European Integration and Environmental Policy*, London: Belhaven Press.

Rüdig, W., Mitchell, J., Chapman, J. and Lowe, P. (1991) 'Social movements and the social sciences in Britain', in D. Rucht (ed.), *Research on Social Movements: the State of the Art in Western Europe and the USA*, London: Westview Press.

Sabatier, P. (1988) 'An advocacy coalition framework of policy change and the role of policy-oriented learning therein', *Policy Sciences* 21, 2–3: 129–68.

Smith, M. (1993) *Pressure, Power and Policy: State Autonomy and Policy Networks in Britain and the United States*, London: Harvester Wheatsheaf.

Standing Advisory Committee on Trunk Road Assessment (1994) *Trunk Roads and the Generation of Traffic*, London: HMSO.

Stokes, G. and Hallett, S. (1992) 'The role of advertising and the car', *Transport Reviews* 12, 2: 171–83.

TEST (1991) *Wrong Side of the Tracks? The Impacts of Road and Rail on the Environment – a Basis for Discussion*, London: TEST.

Ward, H., Samways, D. and Benton, T. (1990) 'Environmental politics and policy', in P. Dunleavy, A. Gamble and G. Peel (eds.), *Developments in British Politics*, London: Macmillan.

3 A 'sustainable' impact on the EU?

An analysis of the making of the Fifth Environmental Action Programme

Annica Kronsell

Toward Sustainability is not a programme for the Commission alone, nor one geared towards environmentalists alone. It provides a framework for a new approach to the environment and to economic and social activity and development, and requires positive will at all levels of the political and corporate spectrums.

(CEC, 1992a: 9)

Introduction

The European Union's Fifth Environmental Action Programme was finally adopted by the Council in February of 1993. The programme is entitled *Toward Sustainability* and will guide the European Union's environmental legislation into the next century. What is unique about it, apart from its sustainability approach,[1] is that it differs radically from previous environmental action programmes. The earlier programmes have been more like shopping lists and formed largely as a reaction to outside events. The Fifth Environmental Action Programme is substantially more detailed and spans more issue areas.

It is a programme based on ideas of ecological modernisation which have originated mainly from discussions within environmental movements. It should create fundamental changes in most sectors of EU policy if it is carried out in its detail. It remains to be seen whether *Towards Sustainability* will succeed in guiding environmental legislation toward a sustainable society where environmental needs are of major concern. Ultimately, this depends on the extent to which the programme can be realised.

The focus here, on the Fifth Environmental Action Programme, should not obscure the fact that environmental questions are still a minor issue area in the EU context. Although the sustainability approach aims at encompassing other sectors and interests through the integration of these sectors into the environmental policies of the EU, economic interests still dominate the EU scene. However, the activity within the EU environmental sector has increased tremendously over the past 20 years.[2] The substantial changes of the late 1980s and early 1990s explain the focus on sustainability in this chapter.

Background

Environmental problems, such as pollution and the extinction of different species of flora and fauna, were observed as early as the 1950s (Jamison *et al.* 1990) and certainly earlier on a local basis (McCormick 1989: 1–24). However, they did not emerge as political issues on the national level of some industrialised countries until the late 1960s and early 1970s. It is generally thought that environmental issues reached the international agenda with the Stockholm Conference of 1972.[3] Until more recently the environment has remained on what is commonly called the low-politics agenda of international relations. By the early 1990s the environment was also discussed in security terms and had moved up from its low-politics status (Thomas 1992; Porter and Brown 1991). This is also true for the EU, as Article 2 of the Maastricht Treaty elevates it to the same status as other issues.

When the EEC was established with the Treaty of Rome of 1957, it was founded as an economic intergovernmental organisation. The Treaty expressed traditional neoclassical economic ideas with customs unions and free trade assumed to result in a higher level of welfare. As co-operation between European nations increased, the policy areas which came to lie under the jurisdiction of the EEC institutions also expanded. When the negative side effects of increased industrialisation and expanding trade between European states became evident, environmental concerns were raised among policy-makers. The First Environmental Action Programme of 1972 was defined as a policy of general concern for European intergovernmental cooperation, but not immediately seen as a matter apt for EEC policy-making (Krämer 1990: 2). Restrained by the economic focus of the organisation, most of the legislation passed in the early stages concerned standardisation and harmonisation measures to eliminate obstacles to free competition between member states (Sands 1991).

With the adoption of the Single European Act (SEA) in 1987 environmental issues gained legal and political recognition in their own right as an explicit European Community goal. When this legal instrument (the SEA) was added to the Treaty, there was no longer any real need for action programmes. Despite this, the heads of state who convened at the Dublin summit in June 1990 expressed a will to continue the tradition of action programmes. The new programme was to be in line with the principles of sustainable development. These principles are also laid down in the Maastricht Treaty, signed in February of 1992.[4]

Aims and focus

This chapter starts from the assumption that political organisations are, generally speaking, stable institutions which contain biases to benefit certain issues and actors in society. The European Community has been working with a strong bias towards economic issues and, particularly, trade relations. Nevertheless, institutions do change and adapt, but in rather unpredictable ways. The claim here is that, with the Fifth Environmental Action Programme, a change has taken place with the acceptance of sustainable development as a policy programme that had been unacceptable earlier. This change can be understood by analysing the ways in which these new ideas, expressed in the programme, were introduced. This programme has emerged through a process whereby certain ideas and certain actors have been particularly influential.

A policy process involves different stages. By the mid-1990s the Fifth Environmental Action Programme was in its implementation phase, but this chapter will be limited to discussing the early stages of the policy process. It focuses on the way changes were introduced onto the EU agenda.

Introducing new issues on an organisational agenda is an interesting part of the policy process, because it is in this phase that the choice of issues available for policy-making are included or excluded. It is fascinating that there are millions of potential issues in society but only a few which emerge as political issues. Theories of agenda setting attempt to shed light on this subject and explain why certain issues become politicised and others do not. Within the broad concept of the environment there are many potential issues – issues ranging from environmental impact assessment and ecotaxes on the one hand, to wildlife habitat protection and ecophilosophy on the other. Some of these reach the political agenda, others do not.[5]

The argument here will be that the temporal sorting (or the garbage can model) can provide a useful framework for analysing the agenda-setting process. This choice is based on the realisation that models which stress strict consequentially and causality (for example bounded rationality, incrementalism) cannot account for the high degree of complexity in the EC policy process (Peters 1994). The claim is, instead, that the temporal sorting model can more adequately explain real events and analyse the empirical material.[6]

The temporal sorting model

The temporal sorting model was developed as a theory about decision-making. Agenda setting is a matter of making a decision about which issues are important. The temporal sorting model views this process as 'a collection of choices looking for problems, [...] solutions looking for issues to which they might be the answer, and decision makers looking for work' (Cohen *et al.* 1972: 2). We can imagine this process as consisting of different streams. They are called streams to convey the idea that they are flowing and in constant motion. They are all important for the policy outcome but the streams develop and operate independently from one another. The streams are linked to each other at certain times and this ultimately depends on how the streams flow (Cohen *et al.* 1976: 27).

I have chosen to speak about three streams, or three separate processes, which develop and influence policy-making separately and independently from one another. The streams are: problems, solutions and participants. When the separate streams come together, policy-making takes place and the outcome is some type of policy, such as the Fifth Environmental Action Programme. The streams come together because the conditions are right: a problem is recognised, a solution is available, and the political climate is right (Kingdon 1984: 92–4).

In the case under scrutiny here we see that the problem stream shows a continuing rapid deterioration of the European environment. In the solution stream some attractive solutions were available. These solutions were acceptable and could appeal to more than the most narrow interests. In the stream of participants there was a strong network of individuals with connections both vertically, down to national and local levels, and horizontally, straddling institutional and national borders. At the same time the political climate and the public opinion were conducive to environmental policies.

The disposition of the analysis will be as follows. First, I will take a close look at the separate streams. Secondly, I will discuss how they come together. Thirdly, I will consider the political climate in which the streams flow.

Problems

Problems can not be described objectively but are a result of individual interpretations of certain facts and events. Environmental conditions come to policy makers' attention, for example, if scientific information alerts them to the fact that there is a problem. Catastrophic events further emphasise an existing condition and put the spotlight on the dangers associated with it.

Scientific evidence

Since different types of governmental or non-governmental bodies monitor societal events, a myriad of indicators can be found around the globe (Kingdon 1984: 95). Indicators are facts, such as disease and death rates, pollution levels and product prices, to mention a few. The indicators do not, in any way, determine whether an actual problem exists but can merely point to a certain condition.

The scientific community has been essential in providing data showing the state of environmental deterioration in various parts of the world. Reports concerning the condition of the environment rely heavily on scientific data. Caldwell (1990: 49) puts it well:

> Mankind's understanding of its place in nature, and an awareness of the predicament in which its efforts to reshape the Earth have placed it, have grown gradually as the concept of the biosphere itself has been built up by successive enlargements of scientific knowledge.

Scientific knowledge is important to EC environmental policy. The EC has produced *State of the Environment* reports showing how serious the environmental problems are in each Member State. One report from June 1992 became part of the scientific base for the drafting of the Fifth Environmental Action Programme (CEC 1992b).

As previously stated, it is not enough to have scientific information showing that we have a problem in a specific environmental area or geographical location. This information must be taken advantage of by a participant. In other words, an individual, such as a policy-maker or an NGO activist, must use this indicator in making a demand for action.

Scientific evidence has been, and is, widely used by policy-makers of all political colours – environmentalists and non-environmentalists alike. Scientific information about a problem is difficult to contradict. The mere official status of such data gives the participant who takes advantage of it, a strong argument without appearing to be political. Or, in the words of an NGO activist: 'It is indeed important that there is a moral or ethical view in environmental policy but it is politically irrelevant. What is essential is rather very strong science that can back up arguments.'[7]

In order to actually illustrate how scientific knowledge influences policy-making, it is more helpful to look at specific legislative measures. The EC passed a car-emission directive which came into effect in 1992. There was plenty of resistance to the directive from the start and a number of different complications occurred (Boehmer-Christiansen and Skea, 1991; Dietz *et al.* 1991). However, one of the Council's arguments against stricter emission control standards was that stricter emission controls were technically and scientifically not feasible. At this point US legislators provided data to the European Parliament Environment Committee showing the newest in emission technology.[8] Simultaneously, the environmental lobby got their hands on the newest technical information on emission controls, provided to them by an ex-official of the US Environmental Protection Agency. A former NGO representative: '[He] knew about the newest and the best. His facts and expertise, which he provided freely and eagerly, were used in discussions with the Commission and in lobbying.'

Scientific knowledge has been very important in providing evidence that environmental conditions have deteriorated. However, it does not say anything about the reasons for a particular condition. It can show us that we have emissions of toxic substances in a particular area and perhaps tell us where the toxic emissions come from, but usually it avoids explaining why we have toxic emissions in the first place.

Events and feedback

Along with other indicators, events or catastrophes can put the spotlight on an issue and, with the help of, for example, the media, demand immediate attention from policy-makers and the public. If a disaster or a crisis is to have such a function, the problem which the event accentuates must already be 'in the back of people's minds' (Kingdon 1984: 103). It is hard to see that any particular event has directly influenced the Fifth Environmental Action Programme, but it has done so in the past on specific initiatives.[9]

Another way of highlighting a condition is through feedback mechanisms and past experiences. The successes and failures of earlier environmental programmes and directives will influence how new programmes are formed.

Experiences with the previous Action Programmes as well as with national-level policies, have led to, as a Commission official puts it, the growing 'realization that governments and their administrations are as much a part of the problem as they are a part of the solution.' Added to this, comes the insight that one Directorate General (DG) cannot realistically be required to solve the problems that the other DGs are creating. The Fifth Environmental Action Programme is very much a reaction to the problems associated with earlier programmes as well as cumulative experiences in the member states.

Solutions

Just as we have indicators pointing to a great number of potential problems in the world, in Europe and in the Member States, we also find ready solutions in the milieu in which policy-making takes place. The most attractive solutions are simple, easily communicated, and easily understood, perhaps with some catch-phrases attached to them. Such neatly packaged solutions are attractive to policy-makers as well as people in general and are more likely to create support and enthusiasm (Olsen and Brunsson 1990: 30–1). The message that the solution contains must appear acceptable, reasonable and in line with the political climate at the time.

The solutions available to the European Commission are developed at national levels in the member states, in non-member states and in international organisations. The solutions which have the greatest impact and the highest probability of getting accepted are the member states' national policies since they are well known and close at hand.

The importance of national examples

On the EU level, national environmental policies are influential and important in the sense that they can exemplify how something can be done. The Dutch National Environmental Plan is a comprehensive plan for environmental management based on the idea of sustainability (VROM *et al.* 1989). This plan was preceded by a study on the Dutch environment (Langeweg 1988) with very dramatic conclusions on the state of the environment. The nature of 'the analysis of environmental problems in [the plan] could have been written by environmental groups such as Greenpeace or Friends of the Earth' (van der Straaten 1992: 51). The Dutch Environmental Policy Plan was a clear precursor to the Fifth Environmental Action Programme. According to an NGO representative: 'If you compare them both, you see they are similar. They both have a thematic approach and they express short-, middle-, and long-term goals.'

The Dutch environmental plan was well known to DG XI and it was in the back of the minds of the core group when they drafted the Fifth Environmental Action Programme. Even if the two plans, the Dutch one and the EU one, are similar in their themes and approaches, they are not identical. Had they been identical, the Dutch plan could not have been accepted as an EU programme. It was important to get a broad consensus around the ideas expressed and this would not be the case had it been perceived to be strictly Dutch. 'Had it been a copy it would have been politically impossible to introduce it as the Fifth Action Programme', said a Commission official.

It was also important that the idea of a more comprehensive plan was well established with the British and the French. There was overall discontent with the previous action programmes, which were more like shopping lists, and a general wish for a more comprehensive framework for environmental policy-making. The Dutch plan was the only complete model which aimed at sustainability.

Sustainability as a solution

Sustainability is the main idea in the Fifth Environmental Action Programme *Toward Sustainability*. Sustainable development is based on the idea that the Earth's resources cannot be exploited indefinitely in accordance with the dominant economic practices of this century. Sustainable development means that development should be consistent with present as well as future needs (WCED, 1987). Thus, growth and development should occur within the carrying capacity of the natural and human environment.

The EU's political leaders committed themselves to the principle of sustainable development at the Dublin summit of June 1990. They undertook to intensify their efforts to protect and enhance the natural environment of the Community itself and of the world of which it is a part. Their aim was to ensure that action by the Community and its Member States would develop on a co-ordinated basis following the principles of sustainable development. Subsequently, the same principle of sustainability was laid down in Article 2 of the Maastricht Treaty signed by all Member States on 7 February 1992. Here sustainability is expressed as being the main objective in respect of future environmental policies. The idea of sustainable development cannot be said to be new, neither with the Fifth Environmental Action Programme nor with the Brundtland report, although the Brundtland report is often cited in this context.[10]

According to an NGO representative the reason appears to be that even though 'the original idea of sustainability came in the International Union for the Conservation of Nature's (IUCN) *World Conservation Strategy* 8 years prior to the Brundtland report, the latter was better packaged, more accessible and therefore more noticed'. At the time of the Brundtland report of 1987 (WCED 1987), sustainability had already been discussed amongst the NGOs for a long time. The concept itself might have been a product of the *World Conservation Strategy* (IUCN, 1980), but O'Riordan, for one, claims that it had been discussed long before this occasion and had been a desirable objective throughout human history (O'Riordan 1988; Adams 1990: 14–41).

The *World Conservation Strategy* originated from a number of IUCN Conferences held in Africa in the 1960s. The central concern was development and conservation of natural resources in the less-developed countries. In this context it was also necessary to consider the human condition and human needs. Some

IUCN ecologists proceeded in the attempt to link development with improvement of the human condition as well as with resource and wildlife protection. The final result was, however, rather superficial: it did not question the fundamental institutional impediments to ecological development on a global basis (O'Riordan 1988: 29–39).

Even though the concept of sustainability originally comes out of a desire from Western environmentalists to consider problems of the developing nations, it also reflects a changing strategy on behalf of the environmental movement. The early environmentalists turned away from the economic and social values of Western high-consumption society, looking for solutions outside the established institutions. In the 1990s many environmentalists turned to a strategy of compromise and mediation. They have proposed solutions which work to incorporate new values into the policies of existing institutions; sustainability is such a strategy. Since the sustainability strategy does not appear to challenge fundamentally the dominant economic growth paradigm, it could be accepted by those that still dominate the EU: economic interests. It is the vagueness and all-embracing nature of the concept of sustainability that has made it acceptable to many political viewpoints, including environmentalists and industrialists alike.

Here we have pointed out that environmental NGOs' ideas have had an impact on the EU institutions. These ideas are not in any way new, neither is it useful or necessary to speak about ideas as new. It is more appropriate to see ideas as something recyclable: old ideas in new packages and in new contexts. One sure point can be made: sustainability is the result of a coupling of green political ideas with classical economic ones.

The participants

Ideas float around 'much as molecules floated around in what biologists call the "primeval soup" before life came into being' (Kingdon 1984: 122–3). As the ideas float around they are aired and discussed by the participants. The ideas sometimes turn into solutions and alternative solutions. In pluralist societies, policies are usually not formed by single individuals, but normally in collectives. Some such groups of participants remain stable for a period of time and can be referred to as networks. Such networks are the settings in which ideas float around and policies are sometimes formed.

Networks

A network approach to interorganisational relations becomes useful in order to understand the relations within issue areas which 'transcend national boundaries and require participation by national as well as international organizations' (Jünsson 1986: 41). The networks extend across the borders of political institutions and are composed of specialists in a given issue area. In the environmental-issue area the specialists are often biologists, zoologists, limnologists, ecologists, social scientists and legal experts.

The idea of networks is, necessarily, both a method and a theory about policy-making. It is a method, because through each individual contact or interview the network exposes itself to the researcher. The theoretical aspect of networks points to connections between a number of individuals who represent different organisations (Thompson *et al.* 1991; Marsh and Rhodes 1992; Jordan and Schubert 1992). It is important to note that these network connections transcend organisational boundaries. Networks played a crucial role in forming the Fifth Environmental Action Programme. In broad terms, one network involved policy-makers and policy initiators in the Commission's DG XI, representatives from the national level, the European Parliament Environment Committee, as well as the environmental organisations active at the European level.

One fundamental aspect conducive to network formation is the fact that individuals often go in and out of different organisations (Cohen *et al.* 1976). Commission officials spoke about the importance of networks in the formation of the Fifth Environmental Action Programme: 'The fact that we had been in and out of the Commission and the Council for many years was essential.' Moving from one organisation to another creates loyalties and friendships across institutional borders. Even though the EU institutions and the lobbying organisations are very different regarding organisational structure, aims, interests, representativeness, responsibility and power, it is very common that individuals are recruited from one to the other.

This going in and out of different Brussels based organisations seems to be rather common. As a consequence, networks develop which transcend the institutional boundaries. The networks are built on personal relationships, often with colleagues at the former workplace. This is not consistent with the structure as it is envisioned by the treaties where the Council is to represent the national interests and the Commission is to work in the European interest together with the European Parliament, which is made up of political groups at the European level. Two policy processes develop: one is the formal process as laid down in the treaties, and the other consists of different networks with individuals not limited by institutional boundaries.

The networks that develop in the EU can be rather complex and have different dimensions with both vertical and horizontal links. The vertical dimension often stems from the individual participant's earlier activities and contacts at the national and local level. Many policy participants have a background in the national or local setting, whether it be in political institutions or non-governmental organisations. The horizontal network is the one created between individuals at the supranational level, in other words between the different EU institutions. Since each participant in the network hypothetically could have extensive webs with contacts in both the vertical and horizontal dimension, a very complex pattern can develop. In such a complex web it would be natural to expect some conflicting interests and loyalties to which the individual is subject.

It is clear that in the case of the Fifth Environmental Action Programme networks played a significant role. As a Commission official made clear, the existence of a well developed network 'is why the Fifth Action Programme has moved quickly and smoothly'. Since the Fifth Environmental Action Programme was heavily influenced by the Dutch National Environmental Plan it was also natural that the key participants in the networks which developed were Dutch.

Cohesion in networks

The frequency of interaction and the degree of cohesion varies between networks. Networks 'are a bit like academic disciplines, each with their own theories, ideas, preoccupations, and fads' (Kingdon 1984: 134). The issues which manage to reach the agenda fit the values that the specialists in the networks have. The specialists in the networks very often perceive the world in similar ways as well as which problems and solutions should be considered (Kingdon 1984: 140–5). The interviews showed that in the network connected to the Fifth Environmental Action Programme, the participants had fairly similar views.

Cognitive factors were also important. In connection with the development of the Fifth Environmental Action Programme one Commission official said:

> A network is extremely important because a certain friendship relationship develops and you can save enormously on time and energy [...] A goodwill develops with mutual respect and integrity. To a person who has been on both sides of the table this experience gives knowledge, yes, but also a lot of respect and contacts.

Various factors influence the relations between individuals in a network and their involvement with a particular issue or issue area. Policy-makers put their efforts and energy into issues which concern, interest and please them (Cohen *et al.* 1976: 31). One important and neglected cognitive aspect in the case of EU policy-making and, perhaps for international networks in general, is the role of language. Language and communication across cultures is crucial to whether the policy-makers can understand and communicate with each other. It is important in the EU context where there are eleven official languages.

It is clear that there was a cohesive network involved in producing the Fifth Environmental Action Programme, but it was a network fairly independent of DG XI. The network extended across the borders of the EU institutions and emerged from those responsible for the programme. It did include those involved in DG XI: 'within the DG XI we have already developed mutual respect and good relationships'. The network was not confined to DG XI. Rather, the consensus obtained for the Fifth Environmental Action Programme was reached because the network included different EU national representatives, especially from the countries which were perceived as obstinate, in this case: Portugal, Spain and Greece. There was at least one other network involved: a vertical network composed of people with similar views and values within the Dutch administration and scientific community. This national network was used by the participants to test the feasibility of the ideas expressed in the programme.

Bringing the streams together

So far this paper has pointed to three streams of processes which have an impact on agenda setting and policy-making. A problem comes up on the agenda and it is coupled with a solution to become policy at certain times when the streams come together. This occurs at a critical time which is not really predictable. Kingdon calls this choice opportunity a policy window and says that a policy window remains open only for a brief time. When a policy window is open, solutions become joined to problems and this creates an opportunity for a participant to push for pet proposals (Kingdon 1984: 173–204). Thus, individuals can be perceived as carriers of pet solutions and problems. Through them the three streams can be coupled or intertwined (Weiner 1976). In the network discussed above, there were a few individuals with a particular interest in seeing the plan through. These participants attached solutions to problems, redrafted proposals in negotiation with other sectors in order to overcome constraints and took advantage of politically propitious events. The policy window was opened because, in the pattern of negotiating action programmes for environmental policy, the time had come for a new programme. The Fourth Environmental Action Programme covered the time period from 1987–92 (Council 1987).

Constraints

So far we have disregarded the formal EU institutions which provide the general framework in which the joining of the streams takes place (March and Olsen 1989). Although the temporal sorting approach has been the starting point here, the formal policy process should not be ignored. The formal process is important because it organises issues into different directorates and units and allocates to different working groups and committees. The selection process takes place according to a praxis established within the EU. The participants, in the process described here, are all more or less working in one of the EU institutions and thus are limited by the rules of procedure and employment policies of those institutions. In DG XI a large percentage of the staff are specialists: highly skilled personnel in particular sectors such as the urban environment, water protection or controlling the environmental impact of products. There is a lack of generalists: staff with a broader view of environmental politics and legislation. This leads to a certain tendency for each staff member to push for his or her own proposals within a narrow field rather than trying to co-ordinate action. The result is many directives in dispersed areas with no clear connection between them. This is the reason why people outside DG XI were called in to deal with the Fifth Environmental Action Programme. As a Commission official explains:

> [The Director General] decided to bring in a group outside of the Commission, a group which could handle relations with lobbyists, national governments, etc. and was not restricted by the organisation of the work in the Commission. There was a need to go outside the Commission structure in

order to do something of the nature which is expressed in the Fifth Action Programme.

Shifts in the administration or other political institutions are likely to open policy windows and create choice opportunities. Shifts within the Commission should be of high relevance, considering that the Commission is the only institution with a formal right of initiative at the EU level. This fact alone suggests that the Commission has important agenda-setting powers. It is not subject to big political shifts since the staff is composed of civil servants and temporary national experts. A reorganisation takes place within the Commission as well as the other institutions when there is a new member state. During the Fifth Environmental Action Programme DG XI was quite stable and not subject to big political shifts. However, the Italian Commissioner Ripa de Meana left toward the end of the process and was replaced with the Greek Commissioner Paleokrassis. Ripa de Meana was very supportive of far-reaching environmental measures[11] and gave DG XI a free hand to carry on with its work. The Commissioner and Director General can be influential regarding which issues are prioritised. The Director General responsible for the Fifth Environmental Action Programme, Mr Brinkhorst, is Dutch and was very much part of the network discussed above.

The Council is a very important institution since it is where policies are approved and final decisions are taken. It is to the Council that the national delegations send their representatives. The Environment Council[12] is involved when environmental issues are considered. This Council can present some major constraints on agenda setting since it, as one Member of the European Parliament (MEP) pointed out,

> is the most powerful [institution] but also the most reactionary. Even the Environment Council is not for the environment but against it. All they do there is come with information from their industry in order to try to find clauses with which they can get exemptions to protect their national industry.

Whether one perceives the Environment Council as reactionary or not depends on who you speak to. Northern European representatives tend to find it reactionary, while southern European delegates, along with the UK and Ireland, tend to find it less so. We can conclude that the Council may act as a constraint since it often expresses a bias based on national interests rather than environmental concerns. This has an impact even at the Commission level, which, to some extent, tries to anticipate the reaction in the Council prior to investing time in drafting policies. The conclusion is that both the Commission and the Council contain some type of bias which restricts the choice of issues. However, since the action programmes are frameworks for future policies, they are more easily accepted by the Council than directives which are binding and have to be implemented at the national level immediately.

The political climate

Whether an opportunity to make choices presents itself or not also depends on whether the political climate is conducive to it. The political climate is important because '[p]eople in and around government sense a national mood' (Kingdon 1984: 154). The policy-makers have a feeling that a rather large number of people are thinking along certain common lines. This mood changes from time to time. If these mood changes are noticed by the policy-makers, it can have important impacts on policy agendas and policy outcomes. The mood is in our case environmentalism, with its origin in the environmental movement.

Characteristic of all of the new social movements[13] is their adherence to a paradigm which contrasts with the dominant structures of the industrial societies. This is based on evidence that post-materialist views have emerged in society.[14] The environmental movement represents such a contrasting view because its approach is based on ecology: an holistic approach which sees human beings in a complex relationship with ecological systems.

Caldwell, for one, sees the environmental movement as a major manifestation of a fundamental transition in human society. This transition has not been fast but its momentum appears to be accelerating (Caldwell 1990: 105). Porter and Brown support this and state: 'The process of paradigm shift has already begun. The sustainable development paradigm has begun to displace the exclusionist paradigm in some multilateral financial institutions, in some state bureaucracies, and in some parliamentary committees [...]' (Porter and Brown 1991: 31). The largest European environmental organisation, the European Environmental Bureau (EEB), argues similarly that there is a general value change in society which is more environmentalist:

> A new paradigm is emerging little by little [...] a new paradigm whereby economics has to go beyond logical monetary values and take into account that any attack against the integrity of the biosphere is an attack on our own integrity.
>
> (EEB 1992a)

Since social movements are dispersed in society they do not have the same type of political impact as organised political forces. Social movements are part of a process. For a time they create a space where new ideas can develop and new interests can be articulated (Eyerman and Jamison 1991: 59–65). If policy-makers perceive that a mood change is taking place, then they will be likely to respond to it promptly (Kingdon 1984: 152–7) and take advantage of the policy window which is opened.

The environmental movement has had such a great impact on Western society that it would be very difficult for a policy-maker to ignore it today. The movement has entered a phase where it has been very influential and it has been institutionalised in many governmental agencies and organisational bodies. The environmental movement has gone through different phases, as all social movements do. Social movements emerge, take an organisational form and become institutionalised or disappear. If the ideas of the social movements are successful in their attempts to affect political and social processes, they change established patterns. If they are less successful, they are discarded (Eyerman and Jamison 1991: 63).

It is difficult to say in which phase the environmental movement is today. Two facts are clear however. One is that it has influenced the European citizens to question their lifestyles and become more aware of the environment of which they are a part. The other fact is that the environmental movement has become highly organised. This is true for most European countries and for the EU as a whole. We have already spoken about the Commission and DG XI, which is responsible for legislative initiatives in the environmental area. The Environment Council has been briefly covered as well. The above institutions are not generally considered environmentalist, even if such ideas can be expressed by individuals working in these organisations.

The environmental movement has responded to the increased activity at the EU level by forming Eurogroups – or NGOs at the European level. Over the past 20 years these groups have expanded their representativeness, with an increased number of national and local organisations as members. They are still a very small percentage of the many Eurogroups at the EU level, with the dominant interest groups being business, industrial and agricultural organisations. Some of the environmental organisations at the EU level are: the EEB, Friends of the Earth Europe, Greenpeace Europe, World Wildlife Fund Europe, Climate Network Europe, Worldwatch Institute Europe, and Environment and Development Resource Center. These Eurogroups are formed with the aim of influencing and monitoring EU policy-making. They are composed of different national, regional and local interest groups.

Generally speaking, among the most important actors outside government are the interest groups. Apart from their ability to block certain types of initiatives, their greatest influence is in preparing alternatives. The Eurogroups are very important in forming alternatives, since the Commission and the European Parliament have comparatively few resources within the institutions to do research, gather facts and prepare proposals for initiatives. They have to rely on expertise from outside the institutions and the European interest groups do fill that role (Nugent 1991: Chapter 3).

Considering this, it is odd to see that when it came to the Fifth Environmental Action Programme, the EU environmental organisations did not have much impact. According to the Commission official responsible for drafting the Fifth Environmental Action Programme, this was due to a lack of interest from the environmental interest groups:

> [T]here was no real reaction from the environmental NGOs. They do a lot of complaining, but when they have the opportunity to present suggestions they don't do it [...] To summarise it: there was no good lobbying – criticism, but no concrete suggestions.

From the perspective of an NGO representative the process is perceived differently: '[T]he policy process has been unnecessarily secretive. I perceive it as top-down in its conception. As [the Fifth Environmental Action Programme] is radical it has not done enough to engage people' (EEB 1992b: 5).

Another example of the organisation of the environmental movement is the Environment Committee of the European Parliament. There are a great number of environmentally-concerned and interested parliamentarians in the Committee. A majority of the MEPs in the Environment Committee consider themselves allies with the environmental interest groups. 'It is undoubtedly an activist Committee', says one MEP. There is a stronger political consensus within the Environment Committee than in any single political group in the European Parliament. It is the second-largest committee and one of the busiest, as it often stages hearings and does report on its own initiatives. The Environment Committee welcomed the Fifth Environmental Action Programme but had some criticisms, whereas the European Parliament felt that the programme did not go far enough. This Committee is very different from any national parliamentary committee since it only has partial legislative powers. Also, it has a strong dedication to 'the promotion of "environmentalism" within EC policy and institutions' (Judge 1992: 209).

Despite the lack of direct influence of environmental movement organisations, environmentalists' ideals were, nevertheless, able to inspire and influence the Fifth Environmental Action Programme in a more indirect way through the network of participants. Many of them have a long record of interest in, and engagement with, environmental NGOs, both at local and national levels. More importantly, these participants adhered to the ideology of ecological modernisation and worked for a reconceptualisation of the relationship between economic issues – the central issue for European Community co-operation – and environmental ones (Weale 1992: 76–9). Ecological modernisation has been the set of ideas adhered to in the drafting phase but also, more recently, in the implementation phase of the programme.

Conclusion

This chapter has described the process through which sustainability reached the EU agenda and became the dominant idea in the Fifth Environmental Action Programme. The process was analysed with the help of a temporal sorting model, which drew attention to three policy streams: problems, solutions and policy-makers. There was a sense of an acute environmental deterioration problem in Europe, which had not been ameliorated by the earlier programmes. There was an attractive solution available and a number of policy-makers were interested in pushing for that solution. These policy-makers were part of a cohesive network that effectively was able to negotiate and generate support for the programme. The temporal sorting model provided a useful tool in looking beneath the formal policy processes and exposing the informal, but transnational and transinstitutional, processes involved in agenda setting.

The ideas originating from the environmental movement were, indirectly, able to influence the Fifth Environmental Action Programme. The drafters, as well as the wider circle of participants, of the programme had past connections with local or national environmental groups and were supporters of the ideology of ecological modernisation. This appeared not to directly challenge the main concern of EU cooperation: economic growth through free trade. Instead, ecological modernisation is a view of environmental protection as a central part of economic growth, which could be accepted and accommodated within the vague and ambiguous notion of sustainability. This had been declared the general policy direction in the Brundtland report (WCED 1987), at the 1990 Dublin EU summit and at the 1992 UNCED conference in Rio. Although sustainable development definitely contains a new direction in environmental policy-making, it remains to be seen whether it will have a forceful and lasting impact on the Community as a whole.

Notes

1 Sustainability means that economic development should be consistent with present as well as future needs. This will be discussed at length further on in the paper.
2 Around 200 environmental directives had been passed by 1995 in the period since the first action programme of 1972.
3 Since the Stockholm conference the environment has been considered an international issue, but it had been discussed in many international fora prior to this.
4 What is actually intended with the concept of sustainability in the Maastricht Treaty is unclear. It is confusing because the Treaty uses the following concepts interchangeably: sustainable progress, sustainable growth and sustainable development. For a more elaborate discussion on this, see Verhoeve *et al.* (1992).
5 In the context of the EU it is also important to add that issues might not reach the EU agenda due to the subsidiarity principle, which means that problems should be solved as close to the citizen as possible. Thus, not every environmental problem in the EU countries is necessarily a matter for EU legislative concerns. This is a matter of conflict and based on a subjective interpretation of a particular problem. The environmental NGOs fear that the renewed emphasis on subsidiarity could be used as a way of repatriating environmental policy or, expressed differently, of making certain that environmental problems remain a national or local concern when they feel they should be resolved at the EU level.
6 Apart from the relevant EU documents and secondary sources, I rely on information gathered through interviews conducted during 1992 and 1993. By using the reputational method or snowballing technique I have been able to pinpoint a number of central actors involved in the making of the Fifth Environmental Action Programme. I have conducted interviews with these actors as well as a number of other individuals closely involved with the EU environmental policy field. These are referred to and quoted from – anonymously – in the text. A majority of those interviewed have stressed that the process is very complex, with many access points and a high degree of coincidence in the policy process.
7 A majority of those interviewed (see note 6) considered scientific evidence as objective and politically neutral.
8 This was provided through the GLOBE network, which is an international organisation of environmentally concerned parlementarians.

9 The toxic storage and spills in Seveso, Italy, led directly to the Seveso directive. The image of the River Rhine as a giant sewer led to directives on rivers in Europe.
10 The Fifth Environmental Action Programme cites it and uses the Brundtland definition of sustainable development (WCED 1987: 8).
11 Carlo Ripa de Meana was the spokesman for the Italian federation of green parties at the time.
12 The Environment Council is made up of the Member States' Environment Ministers and meets two to three times a year. The work in this Council is prepared in working groups where the environmental attachés and other national experts discuss Commission initiatives. The meetings are closed to the public, but the Commission can participate with a representative.
13 There is an ongoing discussion as to whether these movements should be labeled new, or if they are just a variation of old social movements; for such a discussion, see, for example, Morris and Mueller (1992).
14 Both Inglehart (1990); Brand (1990) point to the existence of postmaterialist views in society.

References

Adams, W. M. (1990) *Green Development – Environment and Sustainability in the Third World*, London, New York: Routledge.

Boehmer-Christiansen, S. and Skea, J. (1991) *Acid Politics*, London, New York: Belhaven Press.

Brand, K. W. (1990) 'Cyclical aspects of new social movements: waves of cultural criticism and mobilization cycles of new middle-class radicalism', in R. Dalton and M. Kuechler (eds), *Challenging the Political Order. New Social and Political Movements in Western Democracies*, Cambridge: Polity Press.

Caldwell, L. (1990) *Between Two Worlds – Science, the Environmental Movement and Policy Choice*, Cambridge: Cambridge University Press.

CEC (Commission of the European Comminuties) (1992a) *Towards Sustainability. A European Community Programme of Policy and Action in Relation to the Environment and Sustainable Development*, COM (92) 23 final, Luxembourg: Office for Official Publications of the European Communities.

—— (1992b) *The State of the Environment in the European Community*, Brussels: DG XI.

Cohen, M. D., March, J. G. and Olsen, J. P. (1972) 'A garbage can model of organizational choice', *Administrative Science Quarterly* 17, 1: 1–25.

—— (1976) 'People, problems, solutions and the ambiguity of relevance', in J. G. March and J. P. Olsen *Ambiguity and Choice in Organizations*, Oslo: Universitetsfürlaget.

Council [of the European Communities] (1987) 'Resolution of the Council of the European Communities and the representatives of the governments of the Member States, meeting within the Council of 19 October 1987 on the continuation and implementation of a European Community policy and action programme on the environment (1987–1992)', *Official Journal of the European Communities* 30, C 328 (7.12.1987): 1–44.

Dietz, F. J., van der Straaten, J. and van der Velde, M. (1991) 'The European Common Market and the environment: the case of the emissions of NOx by motorcars', *Review of Political Economy* 3, 1: 62–78.

EEB (European Environmental Bureau) (1992a) *The EU and Sustainable Development –*

Agenda for Community Institutions and Citizens, Brussels: The European Environmental Bureau.

—— (1992b) *Memorandum to the European Council and the United Kingdom Presidency*, Brussels: The European Environmental Bureau.

Eyerman, R. and Jamison, A. (1991) *Social Movements – a Cognitive Approach*, Oxford: Blackwell.

Inglehart, R. (1990) 'Values, ideology, and cognitive mobilization in new social movements', in R. Dalton and M. Kuechler (eds), *Challenging the Political Order. New Social and Political Movements in Western Democracies*, Cambridge: Polity Press.

IUCN (International Union for the Conservation of Nature) (1980) The World Conservation Strategy: Living Resource Conservation for Sustainable Development, Geneva: IUCN.

Jamison, A., Eyerman, R. and Cramer, J. with Laessoe, J. (1990) *The Making of the New Environmental Consciousness: a Comparative Study of the Environmental Movements in Sweden, Denmark and the Netherlands*, Edinburgh: Edinburgh University Press.

Jordan, G. and Schubert, K. (1992) 'A preliminary ordering of policy network labels', *European Journal of Political Research* 21, 1–2: 7–27.

Jünsson, C. (1986) 'Interorganization theory and international organization', *International Studies Quarterly* 30, 1: 39–57.

Judge, D. (1992) 'Predestined to save the Earth: the Environment Committee of the European Parliament', *Environmental Politics* 1, 4: 186–212.

Kingdon, J. (1984) *Agendas, Alternatives and Public Policies*, Boston: Little, Brown & Co.

Krämer, L. (1990) *EEC Treaty and Environmental Protection*, London: Sweet & Maxwell.

Langeweg, F. (ed.) (1988) *Zorgen voor morgen: nationale milieuverkenning 1985–2000*, Alphen aan de Rijn: Samson H.D. Tjeenk Willink.

March, J. G. and Olsen, J. P. (1989) *Rediscovering Institutions – the Organizational Basis of Politics*, New York: The Free Press.

Marsh, D. and Rhodes, R. A. W. (eds) (1992) *Policy Networks in British Government*, Oxford: Clarendon Press.

McCormick, J. (1989) *The Global Environmental Movement*, London: Belhaven Press.

Morris, A. D. and Mueller, C. (eds) (1992) *Frontiers in Social Movement Theory*, New Haven, CT: Yale University Press.

Nugent, N. (1991) *The Government and the Politics of the European Communities*, London: Macmillan.

Olsen, J. P. and Brunsson, N. (1990) 'Introduction', in J. P. Olsen and N. Brunsson (eds), *Makten att Reformera*, Stockholm: Carlssons.

O'Riordan, T. (1988) 'The politics of sustainability', in K. Turner (ed.) *Sustainable Environmental Management – Principles and Practice*, Boulder, Colarado: Westview Press.

Peters, G. B. (1994) 'Agenda-setting in the European Community', *Journal of European Public Policy* 1, 1: 9–26.

Porter, G. and Brown, J. (1991) *Global Environmental Politics*, Boulder, Colarado: Westview Press.

Sands, P. (1991) 'European Community environmental law: the evolution of a regional regime of international environmental protection', *The Yale Law Journal* 100, 8: 2511–23.

van der Straaten, J. (1992) 'The Dutch National Environmental Policy Plan: To Choose or to Lose', *Journal of Environmental Politics* 1, 1: 45–71.

Thomas, C. (1992) *The Environment in International Relations*, London: The Royal Institute of International Affairs.

Thompson, G., Frances, J., Levacic, R. and Mitchell, J. (eds) (1991) *Markets, Hierarchies and Networks: the Coordination of Social Life*, London: Sage.

Verhoeve, B., Bennet, G. and Wilkinson, D. (1992) *Maastricht and the Environment*, London: Institute for European Environmental Policy.

VROM (Ministerie van Volkshuisvesting, Ruimtelijke Ordening en Milieubeheer), Ministerie van Economische Zaken, Ministerie van Landbouw en Visserij and Ministerie van Verkeer en Waterstaat (1989) *Nationaal Milieubeleidsplan: kiezen of verliezen*, 's-Gravenhage: SDU Uitgeverij.

WCED (World Commission on Environment and Development) (1987) *Our Common Future*, Oxford: Oxford University Press.

Weale, A. (1992) *The New Politics of Pollution*, Manchester, New York: Manchester University Press.

Weiner, S. S. (1976) 'Participation, deadlines and choice', in J. G. March and J. P. Olsen (eds), *Ambiguity and Choice in Organizations*, Oslo: Universitetsfürlaget.

4 Ecological modernisation capacity

Finding patterns in the mosaic of case studies [1]

Mikael Skou Andersen

Introduction

Recent years have seen a considerable literature of comparative environmental policy studies evolve, but it is characterised by relatively empirical case studies. Vogel and Kun, observed before the most recent wave of comparative studies, the lack of a more general theory within which to treat the results of the many and diverse case studies that have marked the field. They found that the first generation of environmental policy studies had been fairly descriptive and, while the second generation consisted of somewhat more analytical work, proper third-generation studies, that present or use a comparative theoretical framework, remain relatively rare (Vogel and Kun 1987; Kamieniecki and Sanasarian 1990).

Perhaps the lack of more theoretically based approaches stems from a basic ambiguity about the guiding research question: are we studying environmental policies out of mere curiosity or attachment? Or because governments and research foundations offer ample financial resources for such studies? No, as Martin Jänicke (1990a) has indicated, we are basically studying environmental policy to learn from the relative successes and failures of different environmental policy experiences; or phrased more analytically, we may discern the basic research question of comparative environmental policy studies as one of analysing the ecological modernisation capacity of different political and regulatory systems.

The concept of ecological modernisation has several connotations, but it refers basically to the modernisation of industrial society on the premiss of a more sustainable kind of development (Brunowsky and Wicke 1984; WCED 1987; Fischer 1989; Cairncross 1991; Mol 1995). The old politics of pollution, in which environmental concerns were seen as conflicting with economic development, was superseded by a new politics of pollution, in which environmental concerns were seen as a necessary precondition for continued economic development. The realities of environmental degradation, as well as the economic opportunities associated with the development of environment-friendly technologies, were probably equally important in bringing about this new conceptual understanding of what environmental policy is about. Instead of viewing pollu-tion control as a burden upon the economy, ecological modernisation sees it as a potential source of development. Ecological modernisation theory seeks to explain why different political systems have different capacities for such moder-nisation (Weale 1992: 76).

The concept of ecological modernisation offers, as Weale has pointed out, a set of evaluative criteria for the study of environmental policies. It acknowledges that the links from causes to effects of environmental problems are long and complex, and necessitates a precautionary approach, i.e. prevention rather than restora-tion. It builds on a mass-balance approach according to which the solution of pollution problems by means of dilution or end-of-pipe technology will lead to problem displacement. It does not view pollution control as a sectoral issue, but as an element that needs to become integrated in other policies. In the narrow sense ecological modernisation implies that environmental policy measures need to be source-related and integrated in the production processes, while in a broader sense it addresses the need for structural changes in industrial society (Weale 1992).[2]

Martin Jänicke's ecological modernisation theory (Jänicke (1990a,1995) – see also Jänicke 1978,1990b) – offers one of the few theoretical approaches available to comparative environmental policy studies. It presents a comprehensive but also quite useful framework for the analysis of environmental policy and it offers a set of principal assertions with regard to the conditions for successful ecological modernisation. In this chapter Jänicke's ecological modernisation theory is presented and discussed. The assertion that corporatist political structures should improve the capacity for ecological modernisation is questioned. Furthermore, some methodological and strategic research questions are related to the aggre-gated, country-by-country comparison employed in Jänicke's work and a differ-ent research strategy based on an approach that examines the mosaic of sectoral case studies is both indicated and illustrated.

Jänicke's ecological modernisation capacity theory

Jänicke seeks to explain why some countries are relatively more successful than others in environmental policy. This approach, inspired by the work of Manfred Schmidt (1986), follows the neocorporatist tradition of studies of national economic policy-making. Indeed, Jänicke's work shows that countries with rela-tively successful economic and labour-market policies also tend to have relatively successful environmental policies.

According to ecological modernisation theory (see Figure 4.1), the ecological modernisation capacity of a country refers to its achieved level of institutional and technological problem-solving capabilities. It appears from Figure 4.1 that ecological modernisation capacity depends on four basic variables: (1) problem pressure; (2) consensus ability; (3) innovation capability; and (4) strategy profi-ciency.

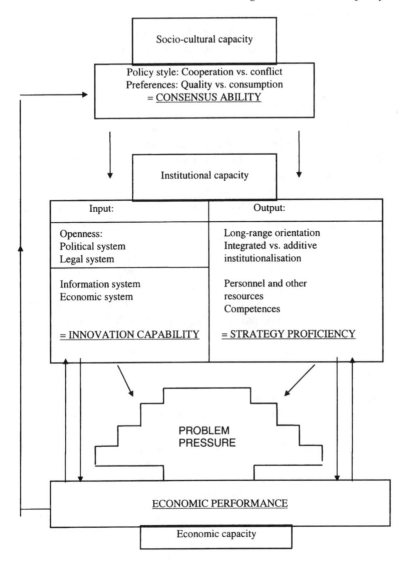

Figure 4.1 Ecological modernisation capacity.
    Sources: Jänicke cited in Andersen 1994:56.

Problem pressure

Problem pressure is mainly a question of economic performance, but the impact of economic performance is two-fold. Countries with adequate economic performance, while willing and able to pay for environmental protection, are also subject to heavier pollution loads. Countries with poor economic performance, which give less priority to pollution control, usually produce less emissions per capita than more affluent countries. The adequate economic performance of a country is, on the one hand, an important precondition for managing environmental problems, but, on the other hand, also for the permanence of pollution.

Policy style (consensus ability)

The national style of policy-making affects the potential for ecological modernisation. Countries where neocorporatist structures prevail are characterised by policy styles more amenable to new interests and ideas, and they promote a more open decision-making process. In such countries environmental issues will, according to the theory, more rapidly become subject to public policies than in countries dominated by closed coalitions that are seeking to safeguard their interests against new challenges. According to Jänicke, the high score on environmental performance in Sweden, the Netherlands and Austria has been reached because of their consensus-seeking approach to decision-making. In Japan the distinctive concertation of interests between industry and the government has assured a comparable and even more successful effort to control pollution. The common denominator of these somewhat different policy styles is the search for broad consensual solutions.

Innovation capability

Institutional innovation capability reflects the policy style as it has become institutionalised over time. Institutional innovation capacity is a property of both state and market institutions, and it determines the degree of openness to innovations and new interests in the political and judicial system, the media, and the economic system. The established constraints or prospects for innovative behaviour help to filter the political and economic inputs. Still, a proper innovation capacity is only a necessary, but not sufficient, condition for successful environmental policy.

Strategy proficiency

Environmental policy will have to be institutionalised in terms of new administrative organisations, programmes, planning, resources and personnel. If institutionalisation is weak – due, for example, to weak administrative authorities or to an environmental policy remaining a sectoral issue rather than a cross-sectional one – the institutional capacity on the output side will be modest. To implement complex policies and programmes is difficult and requires unambiguous competencies and responsibilities. This is especially problematic for federal states, which face potential fragmentation and delay during implementation. If environmental policy is weakly institutionalised, the additional element of environmental bureaucracy may result in similar solutions to end-of-pipe ones, rather than broader, preventive policies.

An assessment

As an analytical model the theory is a complicated tool. In spite of four main variables it offers in fact more than ten variables to investigate. More clear distinctions with regard to what the dependent variable(s) and the independent variable(s) are would be valuable. Still, the framework is quite useful in bringing some order to the many variables that are relevant for comparative studies of environmental policy and it offers a fine starting point for practical analysis. While most of the relevant variables are included in the model, one may question some of the stipulated relations between them, not least the significance attributed to corporatist policy styles.

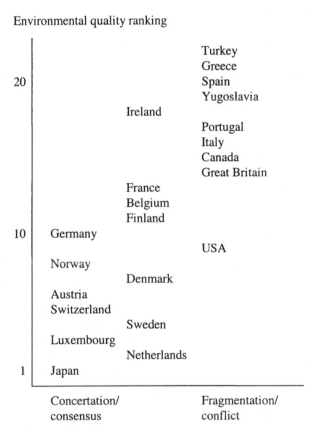

Figure 4.2 A disputable relationship: ranking environmental changes 1970–85, and the integration of interest mediation in western industrial societies. *Sources*: Schmidt 1986; Jänicke 1990a.

Jänicke seeks to validate his theory about the importance of corporatism by ranking the industrialised countries according to environmental performance (see Figure 4.2). Indicators for ecological modernisation capacity are used as a set of environmental indicators, which are pooled to become a single aggregated measure for environmental policy performance. It should be noted, however, that the variables which affect the ranking comprise such heterogeneous parameters as specific air-pollution emissions, the oxygen quality of selected rivers, and sewage-works' capacity. Considerable work has been done to employ data from international statistics for a ranking of industrialised countries, but there are still methodological problems related to this index.

Leaving these problems aside, one must also note that the approach used encounters some difficulties with the ranking of countries with regard to corporatism and interest mediation. In the political science literature there is substantial disagreement about how to rank the industrialised countries according to degrees of corporatism or interest mediation. Lijphart and Crepaz (1991) have identified twelve different ways of ranking the political systems of advanced industrialised countries, which reflects the fact that there is little consensus about how to regard the capabilities and traditions of different countries. It can be noted that the solution of Lijphardt and Crepaz to creating a new index (the Lijphardt–Crepaz index as a simple average of the twelve proposals) underlines the basic ambiguity of this approach. The main problem remains, that there is no authoritative classification of political systems with regard to the degree of interest mediation.

Jänicke does not adhere completely to Schmidt's classification scheme for corporatism (Schmidt 1986: 262), but designs his own. For instance, while Schmidt designates Germany as a consensus-seeking country, but places Sweden and the Netherlands in the middle group, Jänicke classifies Sweden and the Netherlands as consensus seeking and Germany in the middle group. Jänicke's findings indicate a certain relationship between environmental performance and countries with consensus-seeking policy styles, but they are less clear if one, for example, uses Schmidt's classification system. In Figure 4.2 Jänicke's environmental quality ranking (on a scale from 1–20) is shown with Schmidt's classification of interest mediation. It shows that Japan has the highest score on performance, but also that the smaller European states, such as Sweden and the Netherlands, tend to have more successful environmental policies than the larger, federal states, such as Germany and the United States.

In a more recent paper, Jänicke quotes a number of sources which also claim that corporatism is beneficial to the environment (Jänicke, 1995). He could also have mentioned some who have looked for, but not been able to confirm, the importance of consensus seeking. Knoepfel and Weidner, despite their detailed study of air-pollution control in seven European countries, were unable to establish a relation between patterns of consensus seeking and relative success in environmental policy (Knoepfel and Weidner 1985).

The mechanisms of the link suggested between a consensual policy style and environmental policy are substantiated by reference to Katzenstein (1985). The small states of Europe being the forerunners in environmental policy is explained by 'the distinct democratic institutions of these small countries [which] generate stronger political pressure for accommodation and provide better training for an active reform policy' (Jänicke, 1990a: 224, my translation). However, policy determines politics (Lowi 1964; Wilson 1980) and one needs to reflect that the environmental policy sector is quite different from labour-market policies and national economic policies. There are substantial differences between the peak level systems of organisational bargaining that have evolved in the sectors of economic policy-making and in environmental decision-making. The environmental organisations, the counterpart of polluters, tend to be organisationally weaker than the trade unions, and their means of influence are more dependent on the media than on negotiations with the government. Consensus seeking in the environmental policy sector is typically directed from government bureaucracies towards polluters and not often towards the environmental organisations.

Jänicke may have had the German situation in mind. In Germany there is not a tradition for the Ministry of the Environment to include the environmental organisations in negotiations on future legislation (Pehle 1997). Such a tradition exists in the Scandinavian countries, which may thus be seen as a case of accommodation of environmental interests. But does the lack of formal negotiations with environmental interests imply that the ecological modernisation capacity of Germany, all other factors being equal, is less than that of the Scandinavian countries? In many respects environmental policy is stricter in Germany than in the Nordic countries and these countries often look to Germany for specific technical standards (Liefferink and Andersen 1997). Germany has also been more active with regard to the development of EU environmental policy at a high level of protection than, say, Denmark. With the ecological modernisation theory one could of course argue that it is simply due to a higher level of problem pressure in Germany that environmental policies are as strict or perhaps even stricter than in the Nordic countries.

Such an explanation follows obviously from the analytical framework – almost too obviously, one might add. A different explanation, which would challenge the framework somewhat, would be to point to the federal character of the German polity. In a federal context there are more regional interests to accommodate, and consultation between the ministries in Bonn and the German *Länder* is an important axis in the policy-making process (Pehle 1997). There is less room also to accommodate interest groups in the formal negotiations, than in unitary states. We see almost a similar pattern in the EU policy-making process, where the axis in the federal policy-making process runs between the member states and the supranational institutions. Some interest groups are consulted formally, as in an economic and social committee, but they have little real impact. Lobbyists, consulted because of their own initiative, have a much better chance (Andersen and Eliassen 1993).

It is also necessary to add that despite patterns of accommodation in the Nordic countries, environmental organisations have had considerably less influence on pollution-control policies compared with the peak interest organisations of industry and commerce. They are accepted as negotiation partners, but are often excluded from the real bargaining (Andersen and Hansen 1991). From a Nordic point of view the consensual policy style, which puts so much emphasis on reaching agreement with interests (e.g. polluters) before passing new legislation, may also put serious constraints on innovation in environmental policies (Andersen 1994: 76ff) or even take them into a regular deadlock, as has been the case with agri-environmental measures in Denmark and the Netherlands.

Finally a few words about Japan, which has the best score on the environmental performance index and whose political system of concertation is taken as an argument for the importance of interest mediation. Technically, Japan's score on the environmental performance index is explained mainly by the favourable development of the air-pollution parameter. Due to a successful programme relying on economic incentives, Japan's emissions of sulphur dioxide were reduced considerably in the 1970s and are per capita the lowest among the industrialised countries (Weidner 1983; OECD 1991). In other fields of pollution control, notably the water sector, Japan has been doing less well.

The Japanese political system is traditionally not very open. The powerful industrial ministry, the Ministry of International Trade and Industry (MITI), favours concerted action with industry and is not inclined to take other interests into account. This negligence was an important reason why pollution problems reached the level of ecological hara-kiri in the late 1960s. The stringent pollution-control laws that were introduced in subsequent years were not as much a result of the government's propensity to take environmental concerns into consideration, as of the pressure from the so-called seven cities where the political opposition to the ruling Liberal Democratic Party (LDP) was in power. It had created its own expert council to propose environmental measures (Tsuru and Weidner 1985). In fact, Japan used to have in practice, a one-party system in which LDP played the dominating role and LDP also penetrated government bureaucracies. It was rather a temporary departure from this policy tradition – the horrors of Minamata and Yokkaichi acted as catalysts – and was backed up by favourable economic preconditions, which allowed Japan to undertake the successful experiment with the SO_2 pollution tax in the 1970s. During the 1980s less attention was given to environmental policy. The successful achievements of the 1970s were actually a result of a departure from the traditional Japanese policy style, rather than of business as usual (Weidner and Tsuru 1988; Miyamoto 1990; Vogel 1991). They still stand out as rather exceptional, although Japan is regarded as a forerunner in environmental policy (OECD 1994). The Japanese example is complex and it is hardly justified to take it as an argument for the role of concertation and interest mediation in itself. It was hard political conflict which turned the tide in Japanese environmental policies.

Relative success: a question of policy design

Studying modernisation capacity generically, at the level of the nation state and for all sectors of society, raises methodological problems. The problems of constructing valid indicators for generic environmental performance have already been indicated. There are also problems in generalising about the regulatory experiences across different environmental media within the same country. Jänicke's initial research design was in fact more promising. He sought patterns which could explain experiences of *partial* success, and tried to construct a theory on the basis of the patterns found in various case studies:

> Environmentally positive experiences with sector measures are [...] not only to be found in certain front running nations. They are found, though more scattered, in countries with very different overall environmental performances [...] I stress this type of partial success (under otherwise disadvantageous circumstances), because a concept of environmental policy based on the mosaic of partial successes is conceivable.
>
> (Jänicke, 1990a: 214, my translation)

The interesting implication of this approach is that one must then explain the relative success or failure of policies by means of a more specific analysis of regulatory programmes. One needs thus to develop a somewhat more sophisticated understanding of the interplay between the national policy styles and programmatic elements of sector programmes. Not least the role of policy instruments, which is missing in the ecological modernisation capacity theory, needs to be considered.

Jänicke criticises some of the economic policy-instrument theories that investigate the properties of certain policy instruments regardless of the context of their implementation (Baumol and Oates 1988; Jänicke 1995; Andersen 1994,1995b). Still, if a theory about ecological modernisation capacity is about the relative successes and failures of environmental policies, it needs somehow to treat the interplay between the use of policy instruments and the contextual preconditions for their operation. Instrument choice matters as much as institution building.

The reason for the failures of traditional policy-instrument theories has to do with the fact that instrument choice is context bound (Howlett 1991). It is striking that attempts to generate and test hypotheses about the capacities of different policy instruments have not even been able to solve the basic taxonomical problem. There is simply no generally agreed list of the policy instruments available to decision-makers, although many proposals have been made. The taxonomies of policy instruments suggested have varied considerably according to the national background of individual researchers. Lowi's idea, to generate theories about the properties of each type of instrument, has never succeeded (Lowi 1964). Hood (1983); Linder and Peters (1989); Salamon and Lund (1989) have made other taxonomical suggestions which are less complex. However, each of these taxonomies reflect in their own way, as Howlett (1991) has rightly pointed out, the policy styles of particular political systems.[3] The choice of policy instruments reflects unintentionally the institutional background of the different political systems and their specific paths of regulatory development.

Viewed more broadly, each nation's regulatory style is a function of its unique political heritage and each nation regulates the environment in much the same way as it regulates a wide variety of other areas of corporate conduct (Vogel 1986: 128). The term policy style is often used to refer to the distinct national approach to regulation, reflected in settled layers of existing regulations and in new policies that are developed. Policy styles are the standard operating procedures which national political systems have developed for making and implementing policies (Richardson *et al.* 1982: 2). The concept of policy style has proven to be important for the comparative analysis of regulatory policies.[4]

If we reflect on Lundqvist's (1980) classical comparative study *The Hare and the Tortoise*, we may see how differences in policy styles are linked to differences in the polity of different political systems. Similar to the fable, the US started out as the hare with ambitious policy goals, while Sweden more carefully and incrementally, as the tortoise, adopted a gradualistic policy based on what was deemed technologically and politically feasible. The Swedish regulators were more inclined to seek an understanding with industry, rather than using a coercive policy approach like in the US. Whether the tortoise or the hare actually reached targets first was not finally settled, but the difference in policy style is explained by Lundqvist with the differences in the *polity* of Sweden and the US. While the Swedish parliament was dominated by a stable majority (of social democrats) the US Congress was characterised by shifting and competing coalitions. This led Sweden to follow an incremental policy-making process, while in the US Congress issues could become subject to sudden policy escalation. The constitutional difference between the unitary parliamentary Swedish state and the presidential federalism of the US was seen as the ultimate reason for the differences in policy styles.

Vogel (1986) attaches less importance to constitutional and legal factors than to historical and cultural factors for the understanding of national policy styles. On the basis of his comparative study of US and British environmental policy, Vogel concluded that the historical patterns of government–business regulations were generated by a greater willingness of British industrialists to accept increased government controls, by a highly respected civil service and by a general British propensity for cooperation and mutual adaptation (Vogel 1986). Similar observations of the British policy style has been made by others (Hill 1983; Weale 1992). The work of Lundqvist and Vogel may be placed in a broader perspective by the new institutionalism and its distinction between formal and informal institutions (March and Olsen 1989; North 1990). We may see policy styles as having both a formal and an informal element.

A theory of ecological modernisation capacity should basically be able to explain under what circumstances pollution-control programmes will succeed. It will thus have to reveal the interplay between the choice of policy strategy and the most important preconditions for policy-making in terms of the environmental problem pressure and the national style of policy-making. While there are only around thirty advanced industrialised countries, and thus cases to test in the aggregated approach, these countries have produced several hundred environ-

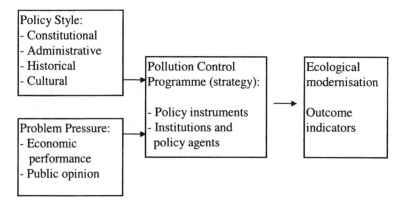

Figure 4.3 Analytical model for the evaluation of environmental policy programmes.

mental policy programmes which offer a rich basis for the testing of hypothesises on ecological modernisation capacity. I urge researchers to compare and analyse these programmes in order to reach a more refined understanding of environmental policy performance and in particular the role that specific policy instruments and policy strategies may play in overall performance. Generalisations on the basis of sectoral experiences may be more helpful in validating the theory about environmental policy capacity than generalisations on the basis of aggregated data and experiences.

I regard Jänicke's analytical framework as a helpful starting point, but in Figure 4.3 I have tried to simplify it and to include the role of policy instruments. The dependent variable is the outcome (relative success or relative failure), while the main independent variable is the pollution-control programme under examination. Furthermore, there are two contextual variables which affect the way that the independent variable functions: problem pressure and national policy style. We want to examine how a particular environmental policy programme and its instruments works under the specific circumstances of problem pressure and policy style. It follows from the question of relative success that this can only be answered by studying and comparing two or more programmes.

National policy styles operationalised

The use of the policy style concept by Vogel and others is unfortunately quite impressionistic. Richardson's formal definition is very narrow and covers only partly what one usually refers to when speaking about national policy styles (compare the case studies in Richardson 1982). Although impressionistic insights used by political scientists may be correct, they only reflect what could be learned by a more detailed study of the constitutional, administrative, historical and cultural institutions and traditions of the particular nation in question. The difficulties which Richardson and Watts had with identifying elements of the national policy style in Germany illustrates the lack of a conceptual methodology (Richardson and Watts 1985).

The *legal-constitutional* system makes up a cornerstone of the national policy style. The historical development of government–business relations takes place within the framework of the legal-constitutional system and the learning process that it generates. The significance of federalism for policy-making can be shown by considering the similarities of US and German policy-making. The gap created by the many administrative levels between policy-makers and the regulated has led policy-makers to impose more stringent and inflexible regulations on industry – a tendency recovered also in the policies of the European Union, a federalist state to be.

The *administrative* set-up, which refers to the division of competencies between state agencies, regional and local authorities, as well as their traditional role in regulatory systems, is a second cornerstone of the national policy style. Among the various administrative levels there is a continuing contest over the distribution of competencies. When a new issue for regulation is specifically put on the agenda, the standard operating procedures for the division of competencies are more open for change. That environmental regulations are delegated to the lowest administrative level does not necessarily imply a low capacity for environmental policy. However, if the local authorities, like in France, are deprived of both competencies and independent sources of income, the set-up will most likely not promote pollution control. To estimate the impact and effectiveness of the administrative arrangements it is also necessary to consider which other interests the various administrative actors may possess. If there is a commonality of interests between regulated industries and local authorities, for instance in securing income and tax revenues, decentralisation may be an inconvenient choice. The establishment or renewal of specific-purpose agencies, like the Dutch district water boards, may promote a more effective administration.

The *historical* and *cultural* elements are often intertwined and are difficult to conceptualise. Historical experiences may have created a certain ideology of public regulation, such as the *Rechtsstaats* ideology of public accountability in post-war Germany (Weale 1992). They may also have supported the development of specific administrative arrangements, such as the Danish farmers movement in the late nineteenth century, which was strongly in favour of decentralisation (Andersen 1995a). In Japan the very notion of the word *kogai* means public health rather than environment and the derived focus on environmental regulation has consequently been one that has neglected pollution problems which are not disastrous to human health (Vogel 1991).

Problem pressure

Problem pressure is the other main contextual variable. When one approaches the issue of environmental policy performance from a sectoral approach, the concept of problem pressure assumes a more specific character. First, problem pressure depends on *basic environmental characteristics*. A high population density or unfavourable geographic conditions may impede the options for dilution, while the opposite may favour them. One of the notable differences between the preconditions for environmental policy in Germany and the UK is that the options for dilution are so poor in Germany.

Secondly, problem pressure depends on *economic performance*, as also pointed out by Jänicke. However, a country with poor economic performance will often display the most visible pollution. Consider the lack of energy efficiency in eastern and southern Europe. In countries with a high GDP per capita the aggregate flow of waste and pollutants will be higher, but less visible, due to the dilution and end-of-pipe technologies. The amount of household waste is twice as high in the Netherlands as in Portugal; while the waste sites are more visible in the South, incinerators produce dioxins and other toxic micropollutants in the North.

Thirdly, problem pressure depends on *public opinion* and the perception of environmental problems among voters. Although opinion polls have shown considerable environmental concern during the last 20 years, it is often only the emergence of green parties that force other parties to adopt environmental issues. Environmental policies develop faster when there is more competition. Without the emergence of *Die Grünen*, Germany would probably not have been such a forerunner in pollution control. Also in the UK, France and Belgium green parties triggered the greening of the political parties. In some countries existing parties may transform themselves into what are more or less green parties, such as D66 in the Netherlands or the Center Party in Sweden. But the impact on other parties is equally important.

The role of the contextual variables

Problem pressure and policy style affect the choice and outcome of pollution-control strategies. A low score on environmental performance indicators may very well be explained by the influence of one or both of these two contextual variables. For instance, if problem pressure is almost absent (in terms of available dilution options, poor economic performance and no public support) it may seriously impede the implementation of any environmental regulation. Furthermore, if the intervention strategy does not match the conditions of the national policy style, implementation may also be impeded. One should be careful in drawing conclusions about the failure of specific policy instruments in countries with difficult preconditions for environmental policy.

The impacts of the policy-style and problem-pressure variables should not be overestimated. Vogel's conclusion (1991) – that Americans have very little to learn from the British approach to pollution control, as each nation's regulatory style is a function of its unique political heritage – is somewhat ethnocentric. Since the industrialised countries face environmental problems which in many respects are similar and in some instances even quite common, they have quite a lot to learn from each other. Although the policy style may bias the functioning of a particular strategy, it should only be regarded as a filter for the implementation of environmental policy programmes.

A classification system for policy designs

To consider the impact of pollution-control programmes, a classification system of different policy designs that occur in this field needs to be established. The usual distinction between command-and-control policies and economic instruments is really too simple to grasp the complex patterns of environmental policy-making. As Royston (1979) noted, there is an American command-and-control policy and a European, more consultative, command-and-control policy. The term command and control stems from the military vocabulary and was used by environmental economists to characterise the regulatory approach in the aftermath of the American Clean Air Act. However, the 1971 Clean Air Act was in many respects rather unique in its emphasis on command and control. It required a reduction of air pollution by 90–95 per cent within less than 3 years and was passed without hardly any formal consultations with automobile producers and other polluters (Marcus 1980). It seems to have had profound implications on the way environmental economists, especially in the US, have come to view environmental policy.

In the western European context there has, with a few exceptions, been a different tradition, based on consultation with the affected interests. The political systems in Europe are different from the fragmented American system and less likely to foster policy escalation. Although the one-sided use of standards and regulations has made it tempting to interpret policies as command and control, the more consultative policy style has frequently resulted in framework regulations open for further negotiations and compromise during implementation. The most interesting exception from this pattern is probably Germany, whose federal political system has led to more of a command-and-control-like policy (see

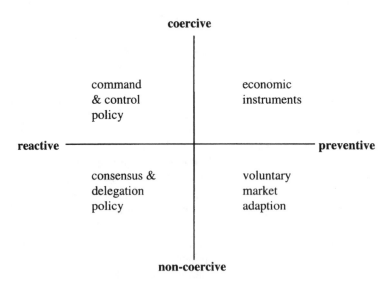

Figure 4.4 Policy designs for environmental policy: a classification system.

below). In other parts of western Europe there has generally been more emphasis on consultation and delegation of responsibility.

The classification proposed in Figure 4.4 operates with four typologies of environmental policy. It indicates on one dimension that environmental policies can be either coercive or non-coercive, a classification which refers to the entire decision-making and implementation process. On the other dimension, policies may be predominantly reactive, which for environmental policies refers to removal strategies of either diffusion or end-of-pipe-solutions. Or alternatively, they can be more anticipatory, in the sense of process-integrated and structural-change policies. From this classification one can deduce four types of environmental policy programmes: command-and-control, consensus-and-delegation, economic instrument, and voluntary market adaption.

In spite of the absence of an authoritative definition, the general understanding of the term command-and-control policy is that command-and-control regulations tend to force all businesses to adopt the same measures and practices for pollution control and thus carry identical shares of the pollution-control burden regardless of their relative impacts. One must differentiate between the use of performance-based and technology-based standards. The latter approach tends to define more narrowly the method and equipment necessary to be in compliance – as, for instance, in terms of a Best Available Technology standard. Performance standards set a uniform control target, but leave some manoeuvring – for example how to meet the target – although its basis will often be a specific technology (Royston 1979; Stavins and Whitehead 1992).

Consensus-and-delegation policies, on the other hand, prescribe no specific *standards* but instead define a *process*, often in a so-called framework law, by which bureaucrats and polluters agree on measures to be taken, either on a case-by-case basis or on the basis of certain agreed-upon performance standards. In its extreme version, bargaining is delegated to local bureaucracies and authorities. Consensus-and-delegation policies have been practised, in varying degrees, in countries such as Denmark, Sweden and the UK. For instance, in the absence of fixed water-emission standards, local authorities in Denmark would give consent to discharges according to the quality of the receiving waters (Lundqvist 1980; Hill 1983; Andersen and Hansen 1991).

Policies based on economic instruments do not prescribe mandatory measures, but establish a system of economic *incentives* by which to charge polluters, either for their emissions, their input of raw materials or their products (Dales 1968; Kneese and Schultze 1975; Ewringmann and Schafhausen 1985; OECD 1989). Whether charges or tradeable pollution offsets are being used, they always leave flexibility to polluters to choose how and to what extent they will reduce pollution. The charging principles may vary, however, from closed tax-bounty schemes to green taxes proper and charging policies often have very different institutional arrangements. Charging policies have been practised in the Dutch and French water policies and in Japanese air-pollution policy.

A fourth possible version of environmental policy is voluntary market adaption (self-regulation), where voluntary agreements, liability schemes, auditing and so on are used to promote ecological modernisation (Cairncross 1991: 236; van Dunné, 1993). The approach is non-coercive because it is based on cooperation with the polluters. It allows for a high degree of flexibility with regard to the technological response and is often more acceptable to the polluters, who may resent the coercive nature of economic instruments. The use of voluntary agreements is being implemented in the 1990s in a number of countries, notably the Netherlands.[5]

The hypothesis suggested here follows the evaluative criteria for ecological modernisation proposed by Weale (see the opening section above). Command-and-control policies as well as consensus-and-delegation policies are suspected of leading to reactive environmental policies. The main reason for the frequent failure of command-and-control policies is that they tend to ignore political and technological constraints. They set up ambitious and ideal targets, which may serve to underline the willingness of policy-makers to do something for the environment, but are unlikely to be met. The most famous example of such failures is probably the US Clean Air Act. The main reason for the failure of the consensus-and-delegation policies is the absence of clear incentives to comply with the ambitions of national policy-makers and the reluctance of local authorities to implement measures costly to local polluters. Pollution can always be exported down the river or over the city border.

Market-based policies are now seen as the most-likely candidates for having the capacity to achieve relative success in promoting ecological modernisation. Thus, economic instruments attempt to include environmental costs in the market economy. Voluntary market adaption builds on mechanisms in the market, like liability, and anticipates the possible self-interest of industries in undertaking ecological modernisation.

However, whether these expectations are justified can only be tested empirically by exploring a number of case studies. Such studies have been rare, especially ones which are not only descriptive but that really seek to evaluate the outcome of environmental policy programmes by means of the comparative approach.

The analytical model illustrated: water pollution and economic instruments

By analysing not only the formal set-up of such pollution-control programmes, but also by considering the impact of policy style and problem pressure, we may be able to discern the impact of the policy strategy from them. How this can be done may be illustrated with a summary of a study on the use of economic instruments for water-pollution control. The following is an extract of the findings which are explained in detail in my book *Governance by Green Taxes* (Andersen 1994). Unfortunately it is not possible within the space of a chapter to convey the real benefits of this approach: these lie very much in the level of details discussed.

The study of water-pollution control investigates the relationship between the policy design of specific national environmental programmes and the outcomes in terms of environmental performance. The focus of interest is the particular background for the choice of specific policies and policy instruments and how such choices affect environmental performance. The goal is to examine how differences in policy designs and the use of policy instruments are reflected in the outcomes of policies. This is done by examining the implementation of the programmes for water-pollution control initiated around 1970 in Denmark, France, Germany and the Netherlands.

Policy designs

As economic instruments have been applied to water policies in Germany, the Netherlands and France since the 1970s, these countries can usefully be compared with Danish water-pollution control policy which did not rely on economic instruments. Are water-pollution control policies marked by similar outcomes in spite of different policies, as implied by Knoepfel *et al.* (1987)? Or have these countries accomplished a relatively more successful environmental policy by having employed economic policy instruments, as claimed by the economists (Baumol and Oates 1988; Kneese and Schultze 1975)?

Danish environmental policy represents a case of consensus-and-delegation policy (Andersen 1989,1995a). No specific emission values were set and the implementation was based on a water-quality principle approach, contracted out from parliament to negotiations between major interest organisations and public officials from the Ministry of the Environment. Also the delegation of considerable discretionary powers to local authorities was another important feature of the Danish Environmental Act. The existence of the conflict-mediating institution, the Environmental Board of Appeal, provided a closed forum for negotiations between industry and bureaucracy on the conditions to be set in the permits, much in line with the principles of corporatism.

Command-and-control has been practised in Germany, where branch guide-lines are based on specific technological solutions: a Best Available Technology approach for most pollutants and a Best Practicable Means standard for organic discharges. In Germany, however, command-and-control policies have been combined with an effluent charge as a supplementary policy instrument. The charge is really an implementation charge, because it is linked to compliance with the standard system.

The French policy design leans more towards the charging approach, due to the use of earmarked charges and specific-purpose agencies: the river-basin agencies. It is distinguished from the German design by the fact that the charge applies to all discharges, although under certain conditions reductions are offered. The absence of national emission guidelines means that the charge takes up an important position as a policy instrument for financing measures by contracts within industries or along rivers. Still, the river-basin agencies remain co-ordinating units, while the responsibility for public sewage works remains with the small and numerous municipalities.

The Dutch policy design is more of a pure charging approach than any of the others. The pollution levy does not only provide marginal funds for pollution control as the French and German levies do, but it constitutes a closed scheme which finances pollution control by industry. Dutch policy-makers have imposed a tax-bounty scheme and arranged a proper institutional framework in terms of special water authorities being responsible for implementation.

Outcome: indicators

From the early 1970s and up to 1987 – the year when Denmark changed its policy design – the outcome has differed remarkably in the four countries.[6] The Netherlands achieved considerable reductions in industrial discharges, whereas Denmark had little success in reducing discharges at the source. In Germany and France there were also reductions, but despite a smaller levy France had achieved more significant reductions than Germany. This is explained with the earmarking of the French levy for water-pollution control subsidies to specific industries through branch contracts. The close relationship between the levies and the emission reductions in the Netherlands is well documented by two Dutch scholars (Bressers, 1988; Schuurman 1988).

Significance of problem pressure

These four countries are among the most affluent in the European Union and have had similar growth rates of GDP. There were some differences in problem pressure, however, related to the geographical conditions for water-pollution control. Most of Germany's rivers are connected in one hydrological basin, with the River Rhine as the major tributary, and conflicts between actors upstream and downstream triggered regulations somewhat earlier than in the other three countries. Denmark had better opportunities for dilution, thanks to the vicinity of the sea to its major cities, while the Netherlands had many canals which were vulnerable to eutrophication. France is more affected by drought than the other three countries, but its industry is mainly located in the northern part of the country. Despite these differences, public support for pollution control has, according to the *Eurobarometer* figures, been relatively strong during the last 20 years. Measured on a uniform European scale, concern about the environment has not been lower in France but there has been more discontent with the lack of measures taken by the government. The pressure was there, but in accordance with the elitist French tradition of government, there was less responsiveness to such sentiments. The anti-nuclear movement in France suffered the same fate. Problem pressure was not the same in the four countries under study, but it was, after all, quite similar in the late 1960s after two decades of industrialisation. The more serious problem pressures in the Netherlands and Germany influenced the policy design – that is the use of economic instruments. But once policies had been designed, they had their own histories of management and effectiveness.

Impact of policy styles on the design and implementation of economic instruments

Denmark

The Danish approach was in accordance with the tradition for broad framework laws that had developed in the wake of corporatism. During implementation, further room was made for negotiations on everything from guidelines to local permits, especially with the influential Federation of Danish Industries. As part of this tradition, administrative courts with representatives from affected interests were usual. To check the powers of the Ministry of the Environment, an independent Environmental Appeal Board was set up to handle complaints on environmental permits. The Appeal Board consists of experts appointed by various interest groups and the Ministry.

There was also a considerable faith in public solutions, in planning, and in the capacity of local authorities. Although there was a certain rhetoric about the polluter pays principle, effluent charges were not introduced. Manufacturing industries were encouraged to discharge to municipal sewage plants and the Environmental Protection Act instructed the municipalities to offer subsidies to the construction of these, causing user fees not to reflect actual costs.

Many public sewage plants were constructed in Denmark during the 1970s and the country became a world leader in this sort of pollution control. Whether the public operation of end-of-pipe solutions would be more expensive than control at the source, or whether public responsibility was to be preferred to private pollution control, was never questioned.

The control of industrial pollution proceeded only slowly, however. Before requirements could be set in relation to manufacturing industries, local authorities had to carry out the tedious process of water-quality planning. This planning was vested in new county authorities, who lacked both the skill and data to carry out the task. In the meantime, local authorities were reluctant to tighten discharge requirements or even to monitor compliance, so as not to lose jobs and tax income. When the counties released the first drafts of water-quality plans, conflicts with municipalities arose over the targets. Consequently, it took more than 10 years before water-quality plans were agreed. In the meantime, there were few incentives for firms to reduce pollution on their own. As a consequence, gross industrial discharges were approximately at the same level in the late 1980s as at the outset, when measured in terms of oxygen-binding substances (see Figure 4.5). About 50 per cent of industrial discharges were treated at public sewage plants, causing reduced pollution; but this was an expensive and essentially problem-displacing solution, as waste water was converted to sludge in increasing volumes. The costs of this approach to water-pollution control was, measured per capita, about four times as high as the Dutch.

The Netherlands

When the Dutch charge system was designed in the mid-1960s there were only

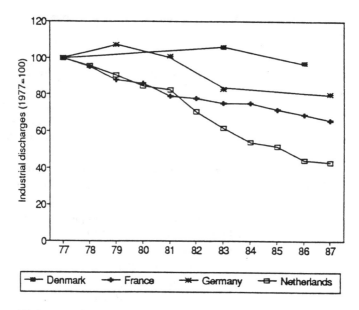

Figure 4.5 Organic discharges from manufacturing industries.
   Source: Andersen 1994: 177.

vague indications of the possible impact of economic incentives on emissions. The use of levies for water-pollution control reflected a century-old tradition of user payment for services from the Dutch district water boards, responsible for dikes, canals and so on. Policy-makers introduced a number of other regulatory instruments for water-pollution control – such as planning, permits and guidelines – much as in other countries. However, these instruments were not as important as the economic instruments, which is shown by just a brief look at the history. The Surface Waters Pollution Act with the charge system came into force in 1970, but the new system of guidelines and permits was not in place before about 1975, because it took time to agree the details. Emissions dropped by about 80 per cent from 1970 to 1989, but the most substantial decline (40 per cent) took place from 1970–5, before the regulatory instruments actually began to work. The Dutch experience cannot be understood without clarifying the role played by the traditional water authorities in its implementation. Rooted in centuries of water regulation, the autonomous water boards, together with the national water authorities, provide a very unusual infrastructure for integrated water management. What made the Netherlands so successful in water-pollution control was essentially that the conventions for water management helped support the establishment of a relatively coercive policy design based on the polluter pays principle.

France

The French policy tradition corresponds to its more centralised system of public administration and has, especially during the Fifth Republic, assumed a somewhat technocratic character reinforced by the weakening of parliament. The French administrative structure is complicated with three levels of local government: *régions*, *départements* and *communes* (municipalities), of which both *régions* and *départements* have a dual character, being led both by a state appointed prefect and an elected council. The *département* is the most important entity, and the traditional Ministries of Industry and Agriculture have branch administrations at the *département* level. A Ministry of the Environment had been established in 1971, but it remained small and insignificant, without its own branch administrations at the *département* level. While the old ministries recruit their officials from each of the traditional engineer corps, the Ministry of the Environment does not refer to such a single administrative corps and lacks the internal and external hegemony that envelopes these corps.

In France, the standard operating procedures for policy-making served to impede the charging programme to some extent, making it somewhat less successful than the Dutch. The French have only partly applied a strategy for integrated water management. The Ministry of Finance has repeatedly interfered and put constraints on the system of environmental charges. And although the so-called river-basin agencies have been established, the Senate and the National Assembly, where the mayors' association has a strong representation, were able to influence the assignment of tasks in the *Loi de l'Eau* so that the local municipalities stayed in charge of water-quality control. France has 36,000 municipalities, which is more than in the rest of the European Community altogether, and in terms of staff and economic resources these are inadequately equipped.

Germany

Germany is a federal state, a fact that penetrates the policy-making process in general. With regard to water pollution, the constitution has delegated competence to the *Länder* level – a disposition that also applied to other pollution-control areas until 1972. For that reason, the federal level has had to work hard to acquire powers concerning the environment.

Germany has a tradition for specific and detailed regulations that often are enforced impeccably. Whether one looks at the German air-pollution regulations or the water-pollution standards, there are often hundreds of specific parameters. The preference for details reflects – apart from the Prussian legacy – also the federal structure in Germany. There has been a long historical development towards the present approach. While in pre-war Germany more than seventy different regional laws regulated discharges to the River Rhine, industry in the 1950s urged the authorities to elaborate similar requirements to make competition equal. In Germany the charging system was less influential and only a flanking instrument to the command-and-control policy. The federal structure and the impending bargaining between the *Länder* and Bonn on administrative responsibilities made it necessary to abandon the high waste-water levy initially proposed, since it could have been blocked by the *Bundesrat*.[7]

Resumé

A major finding of the water-pollution control study is that the design of the economic instruments was influenced in important ways by national policy styles and by institutionalised practices of policy-making. Not only did each of these countries regulate the environment very much according to the way they regulated everything else; the use of economic instruments was also affected by the established, standard operating procedures for applying regulations.

These findings illustrate the difficulties with the sectoral approach if its focus is too narrowly on the use of policy instruments. Policy styles have an impact on policy designs and may limit or enhance the effectiveness of particular policy instruments. Ecological modernisation capacity is thus a question of the institutional preconditions both for designing and implementing a (relatively) successful environmental policy. However, as Lowi (1964) has pointed out policy determines politics, and as the institutional preconditions vary from sector to sector so the ecological modernisation capacity differs between such diverse policy sectors as transport, agriculture and energy policy (Andersen 1995b).

Perspectives

Comparative studies of environmental regulation should be used to help answer our basic research question with regard to the capacity for ecological modernisation. I do not mean this in the sense that a particular type of strategy will always lead to (relatively) successful environmental policy performance. But such studies will improve our understanding of the interplay between policy strategies and the complex political and national environment in which such strategies function.

Everyday officials from environmental ministries travel to other countries to learn from their experience and the development of a common environmental policy within the framework of the European Union represents no less a joint learning process involving fifteen or more countries. Still, in this routine exchange of experiences too little attention is usually devoted to the features of different national approaches to implementation. Ignorance of national deficiencies is a latent condition.

Comparative environmental policy studies may be used to put alleged cases of relatively successful environmental policy under closer scrutiny and to increase the sensitivity towards the particular circumstances under which these have been achieved. The classification of policy designs suggested in this chapter should simply be used to sort out our findings. Given the widespread criticisms of the command-and-control policies of the 1970s, too little evidence has so far been presented to justify the promises of the new politics of pollution.

Notes

1 A previous version of this chapter was published in M. Joas and A.-S. Hermanson (eds.) (1999), *The Nordic Environments: Comparing Political, Administrative and Policy*

Aspects, Aldershot: Ashgate Publishing. Ashgate Publishers kindly granted permission for publication.

2 To use the ecological modernisation capacity concept as the guiding research question does not necessarily presuppose acceptance of the implied relationship between the environment and the industrial economy. Rather, it should lead to a whole set of questions and hypotheses about the preconditions for ecological modernisation, both with regard to the basic structural charcteristics of political systems and their belief systems as with regard to more specific regulatory strategies and the use of policy instruments.

3 Hood's approach reflects unintentionally the British system's preference for consultations and distaste for the use of compulsion. The American approaches reflect the federal and more fragmented tradition prevailing in the US. Lowi's classical taxonomy reflected directly the shift from the use of regulations during the progressive era to the use of expenditure instruments in the Roosevelt period. Canadian policy-instrument theories, e.g. (Doern 1990), mirror more corporatist welfare-state traditions.

4 Richardson *et al.* define policy style as 'the interaction between (a) the government's approach to problemsolving and (b) the relationship between government and other actors in the policy process' (Richardson *et al.* 1982: 13).

5 The Dutch experience has been mixed, as an initial early evaluation of eight covenants showed (Klok 1989).

6 For more indicators, also on economic effectiveness, see Andersen (1994).

7 The *Bundesrat* is the second chamber and consists of delegates from the *Länder* governments.

References

Andersen, M. S. (1989) 'Miljøbeskyttelse – et implementeringsproblem', *Politica* 21, 3: 312–28.

— (1994) *Governance by Green Taxes: Making Pollution Prevention Pay*, Manchester, New York: Manchester University Press.

— (1995a) 'Conflict, compromise and capacity: the ecological modernisation process in Denmark', paper prepared for the WIDER workshop on 'National Environmental Policies – A Comparative Study of Capacity Building', Wissenschafszentrum, Berlin, 5–6 May.

— (1995b) The importance of institutions in the design and implementation of economic instruments in environmental policy, ENV/EPOC/GEEI(95)2, Paris: OECD Environment Directorate.

Andersen, M. S. and Hansen, M. W. (1991) *Vandmiljo&z.shtsls;planen: fra forhandling til symbol*, Harlev J.: Niche.

Andersen, S. S. and Eliassen, K. A. (eds) (1993) *Making Policy in Europe. The Europeification of National Policy-Making*, London: Sage.

Baumol, W. J. and Oates, W. E. (1988) *The Theory of Environmental Policy*, Cambridge: Cambridge University Press.

Bressers, H. (1988) 'A comparison of the effectiveness of incentives and directives: the case of the Dutch water quality policy', *Policy Studies Review* 7, 3: 500–18.

Brunowsky, R. and Wicke, L. (1984) Der Öko-Plan. Durch Umweltschutz zum neuen Wirtschaftswunder, München: Piper.

Cairncross, F. (1991) *Costing the Earth*, London: Business Books.

Dales, J. H. (1968) *Pollution, Property and Prices*, Toronto: University of Toronto Press.

Doern, G. B. (ed.) (1990) *Getting it Green: Case Studies in Canadian Environmental Regulation*, Toronto: C. D. Howe Institute.

Ewringmann, D. and Schafhausen, F. (1985) *Umweltbundesamt Berichte 8/85: Abgaben als ökonomischer Hebel in der Umweltpolitik*, Berlin: Erik Schmidt Verlag.

Fischer, J. (1989) *Der Umbau der Industriegesellschaft*, Frankfurt: Eichborn Verlag.

Hill, M. (1983) 'The role of the British Alkali and Clean Air Inspectorate in air pollution control', in P. B. Downing and K. Hanf (eds.), *International Comparisons in Implementing Pollution Laws*, Boston, MA: Kluwer–Nijhoff.

Hood, C. (1983) *The Tools of Government*, London: Macmillan.

Howlett, M. (1991) 'Policy instruments, policy styles, and policy implementation: national approaches to theories of instrument choice', *Policy Studies Journal* 19, 2: 1–21.

Jänicke, M. (1978) *Umweltpolitik: Beiträge zur Politologie des Umweltschutzes*, Opladen: Leske & Budrich.

—— (1990a) 'Erfolgsbedingungen von Umweltpolitik im Internationalen Vergleich', *Zeitschrift für Umweltpolitik*, 3/90: 213–31.

—— (1990b) *State Failure – the Impotence of Politics in Industrial Society*, Cambridge: Polity Press.

—— (1995) 'The political system's capacity for environmental policy', paper prepared for the WIDER workshop 'National Environmental Policies – A Study of Capacity Building', Wissenschaftszentrum, Berlin, 5–6 May.

Kamieniecki, S. and Sanasarian, E. (1990) 'Conducting comparative research on environmental policy', *Natural Resources Journal* 30, 2: 321–40.

Katzenstein, P. (1985) *Small States in World Markets*, Ithaca, NY: Cornell University Press.

Klok, P.-J. (1989) *Convenanten als instrument van milieubeleid*, Enschede: Twente Universiteit.

Kneese, A. V. and Schultze, C. L. (1975) *Pollution, Prices and Public Policy*, Washington, DC: Brookings Institute.

Knoepfel, P. and Weidner, H. (1985) *Luftreinhaltepolitik im internationalen Vergleich*, Vol. 1–6, Berlin: Edition Sigma.

Knoepfel, P. Lundquist, L., Prud'homme, R. and Wagner, P. (1987) 'Comparing environmental policies: different styles, similar content', in M. Dierkes, H. N. Weiler and A. B. Antal (eds), *Comparative Policy Research. Learning from Experience*, Aldershot: Gower.

Liefferink, D. and Andersen, M. S. (eds) (1997) *European Environmental Policy. The Pioneers*, Manchester: Manchester University Press.

Lijphart, A. and Crepaz, M. M. L. (1991) 'Corporatism and consensus democracy in eighteen countries: conceptual and empirical linkages', *British Journal of Political Science* 21, 2: 35–58.

Linder, S. H. and Peters, G. (1989) 'Instruments of government: perceptions and contexts', *Journal of Public Policy*, 9, 1: 35–58.

Lowi, T. (1964) 'American business, public policy and political theory', *World Politics* 16, 4: 677–715.

Lundqvist, L. J. (1980) *The Hare and the Tortoise: Clean Air Policies in the United States and Sweden*, Ann Arbor, MI: University of Michigan Press.

March, J. G. and Olsen, J. P. (1989) *Rediscovering Institutions*, New York: The Free Press.

Marcus, A. A. (1980) *Promise and Performance: Choosing and Implementing an Environmental Policy*, Hartford, CT: Greenwood Press.

Miyamoto, K. (1990) 'Environmental problems and environmental policy in Japan after the second world war', in DocTer International Institute for Environmental Studies (ed.) *European Environmental Yearbook*, London: DocTer International Institute for Environmental Studies.

Mol, A. (1995) *The Refinement of Production: Ecological Modernization Theory and the Chemical Industry*, Utrecht: van Arkel.

North, D. C. (1990) *Institutions, Institutional Change, and Economic Performance*, Cambridge: Cambridge University Press.

OECD (1989) *Economic Instruments for Environmental Protection*, Paris: OECD.

—— (1991) *Environmental Indicators*, Paris: OECD.

—— (1994) *Environmental Performance Reviews: Japan*, Paris: OECD.

Pehle, H. (1997) 'Domestic and foreign environmental policy in Germany', in D. Liefferink and M. S. Andersen (eds.), *European Environmental Policy: the Pioneers*, Manchester; Manchester University Press.

Richardson, J. (ed.) (1982) *Policy Styles in Western Europe*, London: George Allen & Unwin.

Richardson, J. and Watts, N. (1985) *National Policy Styles and the Environment: Britain and West Germany Compared*, Internationale Institut für Umwelt und Gesellschaft (IIUG) dp 85–16, Berlin: WZB.

Richardson, J, Gustafsson, G. and Jordan, G. (1982) 'The concept of policy style', in J. Richardson (ed.) *Policy Styles in Western Europe*, London: George Allen & Unwin.

Royston, M. G. (1979) *Pollution Prevention Pays*, Oxford: Pergamon Press.

Salamon, L. M. and Lund, M. S. (1989) 'The tools approach: basic analytics', in L. M. Salamon (ed.) *Beyond Privatization: the Tools of Government Action*, New York: The Urban Institute Press.

Schmidt, M. (1986) 'Politische Bedingungen erfolgreicher Wirtschaftspolitik', *Journal für Sozialforschung* 26, 3: 251–73.

Schuurman, J. (1988) *De prijs van water*, Arnhem: Gouda Quint.

Stavins, R. N. and Whitehead, B. W. (1992) 'Dealing with pollution', *Environment* 34, 7: 7–41.

Tsuru, S. and Weidner, H. (1985) *Ein Modell für uns: die Erfolge der japanischen Umweltpolitik*, Köln: Kiepenheuer & Witsch.

van Dunné, J. M. (1993) Environmental Contracts and Covenants: New Instruments for a Realistic Environmental Policy, Lelystad: Vermande.

Vogel, D. (1986) *National Styles of Regulation: Environmental Policy in Great Britain and the United States*, Ithaca, NY: Cornell University Press.

—— (1991) 'Environmental policy in Europe and Japan', in N. J. Vig and M. Kraft (eds.), *Environmental Policy in the 1990s*, Washington, DC: CQ Press.

Vogel, D. and Kun, V. (1987) 'The comparative study of environmental policy: a review of the literature', in M. Dierkes, H. N. Weiler and A. B. Antal (eds), *Comparative Policy Research*, Aldershot: Gower.

WCED (World Commission on Environment and Development) (1987) *Our Common Future*, Oxford: Oxford University Press.

Weale, A. (1992) *The New Politics of Pollution*, Manchester, New York: Manchester University Press.

Weidner, H. (1983) 'Luftreinhaltepolitik in Japan', *Zeitschrift für Umweltpolitik* 3, 2: 211–47.

Weidner, H. and Tsuru, S. (eds) (1988) *Environmental Policy in Japan*, Berlin: Sigma.

Wilson, J. Q. (1980) *The Politics of Regulation*, New York: Basic Books.

5 Structural change and environmental policy

Martin Jänicke, Harald Mönch and Manfred Binder

Two fundamental strategies can be distinguished in environmental policy: a remedial end-of-pipe strategy that does not revise existing problematic technologies and a preventive strategy that alters production and consumption *ex ante* towards ecologically better adapted forms (Gerau 1978; Jänicke 1979; Simonis 1989). A similar distinction can be drawn between an add-on strategy, as in end-of-pipe approaches, and an integrated strategy based on environmental protection. The term preventive, or integrated, environmental-protection policy as we understand it, refers to all forms of environmental policy or ecologically motivated economic policy that envisage structural change of the economy. Technological progress in the sense of ecological modernisation forms part and parcel of this concept of structural change. It is after all part of the process – covered nowadays by a variety of terms – of the transition to sustainable development or eco-restructuring. We refer to the pursuit of such aims as ecological structural policy. Integrated environmental protection of this sort requires not only an integrated technical solution, but also an integrated political solution in the sense of well-developed cooperation between policy fields.

After some comments on the concepts of structural environmental stress and ecological structural policy, we will present the main results of a study on ecologically important structural change in the industrialised economies during the 1970s and 1980s. Four aspects of this change will be dealt with in greater detail: energy prices, neglected technological possibilities for ecological modernisation, the importance of intra- and intersectoral change in the manufacturing sector, and the low significance attached to sectoral structural policy. A few cautious policy recommendations will conclude the chapter.

We call the object of ecological structural policy structural environmental stress. Every economic practice causes structural environmental stress in so far as, without additional cleaning technology, it would cause actual environmental damage. The extent of structural environmental stress, therefore, is not influenced by the end-of-pipe measures being taken. As a rule, structural environmental stress involves considerable residual pollution, even where remedial environmental protection measures are taken. In the case of a coal-fired power station, for example, end-of-pipe cleaning technologies do nothing to eliminate the environmental problems that remain. Some of these are even caused by the cleaning-up process. The residual problems include CO_2 emissions; residual emissions, which as a result of growth processes can increase further; extremely high water consumption; heavy transport commitment (fuels, limestone, gypsum and other waste); high waste production; waste-water production from cleaning plants; the detrimental impact on the landscape from coal extraction, limestone quarrying or power lines; considerable consumption of electricity by cleaning plants; and the consumption of resources. Environmental problems are often simply shifted from one region or medium to another by end-of-pipe environmental protection measures. Moreover, remedial environmental protection presupposes goods production that is detrimental to the environment.

Environmental protection, especially in the long-term, will therefore necessarily be conceived in terms of structural policy. This is inevitable, not least of all because residual problems can rapidly return to their initial levels due to growth processes. This may also be formulated as an ecological–economic imperative: in a growing economy the ecologically detrimental impact of growth must constantly be compensated for. Initially, remedial environmental technology appears appropriate. When this strategy has been exhausted, stress-reducing technological and structural changes become essential. A critical point is whether environmental policy can overcome its status as mere add-on policy and be integrated into other policy fields. In this case the point of gravity in environmental policy would shift to other policy areas like energy, agricultural, transport or industrial policy. This chapter deals with industrial policy (Johnson 1984) as ecological structural policy. This includes all measures designed to reduce structural environmental stress through a different approach to industrial production.

Environmental protection as a sectoral restructuring policy may seek structural change in the sense of a change in the relative status of individual sectors: intersectoral change. The change may be tackled politically by means of ecologically motivated shutdowns, branch agreements, environmental charges and so on. It may, however, also be the result of changes in demand – because of, for example, changes in values or of modernisation processes among industrial customers. Demand may change because investors switch to other branches or relocate production, which amounts to shifting ecological problems. Shifting freight transportation from roads to railways can also be regarded as sectoral structural policy.

Ecological structural policy may, however, also envisage intrasectoral change. In this case it is generally a matter of technological innovation in the sense of ecological modernisation (Jänicke 1984; Zimmermann *et al.* 1990; Garnreiter *et al.* 1986). Ecological modernisation as a programme recognises that, in view of the problem of long-term destruction of the environment, technological progress should not be retarded, but rather accelerated in order to achieve the switch to ecologically more-appropriate production and products. Ecological modernisation involves both innovation in production processes and innovation in products.

Changes in products and processes under the heading of soft chemistry have been discussed and also implemented. In the field of electricity supplies, for example, it is possible to reduce environmental stress via highly-efficient power stations and combined heat and power production or via the transition to cleaner energy sources like water, gas or solar energy. In the building industry, the transition to ecological building materials and building forms, like zero-energy buildings with low space and water needs, represent a similar possibility. Intra-sectorally-oriented industrial policy will generally be determined by the aim of lowering intensity (per unit of value added) for special parameters of production (Jänicke 1984; Fischer-Kowalski 1991; Ayres 1992; Lucas *et al.* 1992; Schmidt-Bleek 1994). This particularly affects material intensity (especially with non-renewable resources), energy intensity, water intensity, land-use intensity, trans-portation intensity, emission intensity, waste intensity and risk intensity.

In the case of intrasectoral structural improvement, the organisational frame-work and the social reality of the sector remain essentially unchanged. Change affects the form and content of production. For employment-policy or regional-policy reasons it will thus frequently seem best to practise environmental struc-tural policy initially in this innovative variant. Much can be said in favour of inducing the process of ecologically-beneficial intrasectoral change primarily by means of technological progress within sectors of the economy. On the one hand, ecological modernisation results in a high demand for tertiary pre-produc-tion activities like research, development, consultancy, and educational qualifi-cation. This increases the importance of services that have little impact on the environment. On the other hand, production sparing in energy and materials leads to an (at least relative) decline in basic industries causing environmental stress.

The overall impact of shifts between sectors – intersectoral change – may be to reduce environmental stress, but this may also disguise problem relocation in the form of adjustments in the division of labour in the world market (Low 1992; Stevens 1993; DIW/RWI 1993). Technological change as ecological modernisa-tion, by contrast, tackles the problem directly.

Comparison of changes in structural environmental stress in the industrialised countries

How has structural environmental stress changed in the industrialised countries since 1970? The following treatment of this topic is based on an international comparative research project at the Free University of Berlin (Jänicke *et al.* 1992 and 1997).[1] A total of thirty-two industrialised countries were examined for the period 1970–90 with regard to eleven products that produce a particularly high degree of structural environmental stress: crude steel, aluminium, copper, lead, zinc, cement, petroleum products, chlorine, pesticides, fertilisers and paper/paperboard. These products are representative of the basic industries most detri-mental to the environment: paper and paperboard industries, basic metal indus-tries, chemical industries and non-metallic mineral industries. In West Germany they represented 73 per cent of energy consumption, 83 per cent of water consumption, 49 per cent of solid waste generation (76 per cent of hazardous waste) and 60 per cent of all investment in pollution abatement and control of the manufacturing industries by the end of the 1980s; their combined value-added amounted only to 21 per cent of manufacturing (see also Lucas *et al.* 1992). We added two non-manufacturing sectors – electricity generation and freight transportation – which can also serve as background variables of structural environmental stress. Of course, this analysis does not allow any kind of environmental impact assessment of national economies: it is mostly confined to industrial production and does not deal with post-consumption pollution, for example cars. Furthermore, it ignores any end-of-pipe measures and differences in the carrying capacity of the environment in different countries.

The research showed that advanced industrialised countries in traditional sectors such as cement, crude steel, petroleum products (to a small extent) or fertiliser achieved at least relative reduction in structural environmental stress by uncoupling these production sectors from overall economic growth (Jänicke *et al.* 1997). The same is true for most of the non-ferrous metals (with the exception of aluminium, see below). A trend towards uncoupling energy requirements and the weight of freight moved by road and rail was also observed. These changes cannot be explained by the relocation of production to Third World countries, because consumption data show similar patterns. The highly-industrialised economies are net exporters of most of the materials discussed here, with the notable exception of several non-ferrous metals and (to a small extent) petroleum products (Jänicke *et al.* 1997). The less-advanced industrial countries in eastern Europe showed further deterioration from an already high level in the sectors mentioned. The greatest degree of structural deterioration, albeit at a low initial level, was evidenced by southern European countries (see Figures 5.1 and 5.2).

By contrast, structural environmental stress from electricity generation and freight movement by road, as measured in tonnes per km, increased on average in ways linked to rates of economic growth in most of the advanced industrialised countries. No uncoupling has occurred in this respect; instead, frequent increases to higher levels. In a number of advanced industrialised countries this is also true for primary aluminium production, chlorine, paper and paperboard and (where figures are available) pesticides (see Figures 5.3 and 5.4).

The significance of structural environmental stress reduction in traditional industries and the increase in more modern sectors is well illustrated by the example of the Federal Republic of Germany (FRG) (see Figure 5.5). The two breaking points of the first and second oil crises (1973 and 1979) are particularly apparent.

Four aspects of this change need to be dealt with in greater detail: the significance of the energy prices, particularly industrial electricity prices; the contrast between technological capability and its actual utilisation, especially in relation to energy and transport; the considerable significance of technological (intrasectoral) change and the low significance of intersectoral change; and the apparently low significance of sectoral structural policy for ecologically beneficial development.

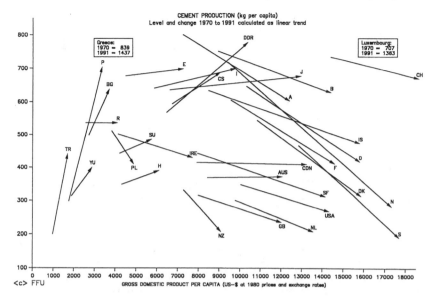

CEMENT PRODUCTION (kg per capita)
Level and change 1970 to 1991 calculated as linear trend

Figure 5.1 International change profile of cement production 1970–91.
Abbreviations: A, Austria; AUS, Australia; B, Belgium; BG, Bulgaria; CDN, Canada; CH, Switzerland; CS, Czechoslovakia; D, West Germany; DDR, East Germany; DK, Denmark; E, Spain; F, France; GB, Great Britain; GR, Greece; H, Hungary; I, Italy; IRE, Ireland; IS, Iceland; J, Japan; L, Luxembourg; N, Norway; NL, Netherlands; NZ, New Zealand; P, Portugal; PL, Poland; R, Rumania; S, Sweden; SF, Finland; SU, Soviet Union; TR, Turkey; USA, United States of America; YU, Yugoslavia.

The significance of energy prices

In northern Europe (Scandinavia and Iceland), North America (USA and Canada) and Oceania (Australia and New Zealand) – countries with particularly low electricity prices – energy requirements developed to a high level with untypical dynamism. This was especially true for electricity consumption: where a country charged electricity prices twice those prevailing in a second country, the former generally showed a more intense use of electricity at a rate a third lower (see Figure 5.6). These electricity-intensive countries also have industries causing a particularly high degree of environmental stress (aluminium, chlorine, paper and paperboard) – a double-negative structural impact. It is thus hardly by chance that Japan, the country with the highest industrial electricity prices, also shows the highest degree of structural change.

Especially with regard to the discussion on the ecological impact of energy taxes, however, the massive price differentials necessary to obtain this result should not be overlooked. If the long-term relation between electricity price and electricity consumption, established in international comparative terms, were to exist also for short-term elasticity within individual economies, a country

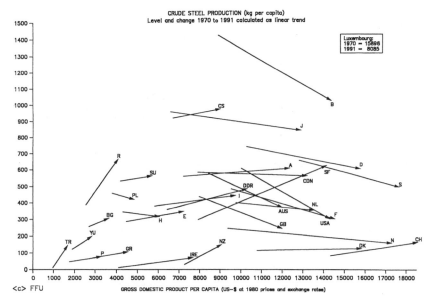

Figure 5.2   International change profile of crude steel production 1970–91.
Abbreviations: A, Austria; AUS, Australia; B, Belgium; BG, Bulgaria;
CDN, Canada; CH, Switzerland; CS, Czechoslovakia; D, West Germany;
DDR, East Germany; DK, Denmark; E, Spain; F, France; GB, Great Brit-
ain; GR, Greece; H, Hungary; I, Italy; IRE, Ireland; IS, Iceland; J, Japan;
L, Luxembourg; N, Norway; NL, Netherlands; NZ, New Zealand; P,
Portugal; PL, Poland; R, Rumania; S, Sweden; SF, Finland; SU, Soviet
Union; TR, Turkey; USA, United States of America; YU, Yugoslavia.

with an average annual growth rate of 3 per cent would need to double relative
electricity prices every 16 years simply in order to keep electricity consumption
constant. Persistent political intervention in price structures on this scale would
doubtless be a politically fraught undertaking.

Unexploited technological possibilities

Energy requirements in the more advanced industrialised countries generally
divert from the economic growth trend. Some of these countries evidence stagnation in per capita consumption of primary energy. The United States,
Denmark, the UK and Luxembourg even experienced a decrease. The majority
of the more advanced industrialised countries have at least exhausted some of the
conservation potential, which today can be regarded as the state of the art in
energy research. But there is no development towards stagnation in electricity
consumption parallel to that registered with energy requirements. In the course of
time and with the rise in the standard of living there has hitherto always been an
increase in electricity generation.

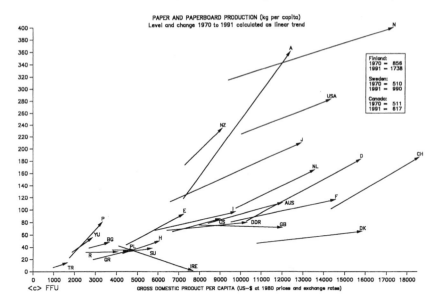

PAPER AND PAPERBOARD PRODUCTION (kg per capita)
Level and change 1970 to 1991 calculated as linear trend

Finland:
1970 = 856
1991 = 1738

Sweden:
1970 = 510
1991 = 990

Canada:
1970 = 511
1991 = 817

<c> FFU

GROSS DOMESTIC PRODUCT PER CAPITA (US-$ at 1980 prices and exchange rates)

Figure 5.3   International change profile of paper and paperboard production 1970–91.
Abbreviations: A, Austria; AUS, Australia; B, Belgium; BG, Bulgaria;
CDN, Canada; CH, Switzerland; CS, Czechoslovakia; D, West Germany;
DDR, East Germany; DK, Denmark; E, Spain; F, France; GB, Great Brit-
ain; GR, Greece; H, Hungary; I, Italy; IRE, Ireland; IS, Iceland; J, Japan; L,
Luxembourg; N, Norway; NL, Netherlands; NZ, New Zealand; P, Portu-
gal; PL, Poland; R, Rumania; S, Sweden; SF, Finland; SU, Soviet Union;
TR, Turkey; USA, United States of America; YU, Yugoslavia.

Some affluent industrialised countries (Norway, Canada, Iceland, Sweden)
even exhibited an accelerated increase. As mentioned above, the countries
concerned charge low electricity prices. In Sweden, for example, the advance
of electricity into the heating market took place – a questionable development
from the energy policy point of view. Because of the particularly large volume of
waste heat produced by electricity generation, this strong trend towards electricity use was in conspicuous contradiction to the objective of conserving primary
energy. In other words, if the move away from oil policy had been less a policy
emphasising electricity, a great deal more primary energy would have been
saved. However, awareness of significant technological conservation capacities
in final electricity consumption entered public awareness only in the 1980s.

In the field of freight movement the contrast between technological possibilities and reality was no less marked. If we consider the weight of cargo moved by
rail and road, we find a trend towards stagnation in a series of advanced industrialised countries, in contrast to eastern Europe with its then prevailing tonnage
ideology. Sweden, Japan, France, the UK, Belgium and Luxembourg even
showed a per capita reduction. This remarkable result seems to indicate that

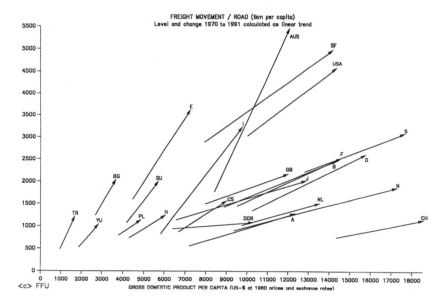

Figure 5.4   International change profile of freight movement 1970–91.
Abbreviations: A, Austria; AUS, Australia; B, Belgium; BG, Bulgaria; CDN, Canada; CH, Switzerland; CS, Czechoslovakia; D, West Germany; DDR, East Germany; DK, Denmark; E, Spain; F, France; GB, Great Britain; GR, Greece; H, Hungary; I, Italy; IRE, Ireland; IS, Iceland; J, Japan; L, Luxembourg; N, Norway; NL, Netherlands; NZ, New Zealand; P, Portugal; PL, Poland; R, Rumania; S, Sweden; SF, Finland; SU, Soviet Union; TR, Turkey; USA, United States of America; YU, Yugoslavia.

growth has incorporated a more distinctly qualitative component. The fact that the consumption of raw materials in the FRG in 1990 was slightly below the level for 1978 (see below) is fully compatible with stagnation in the weight of cargo moved. This is, however, in strong contrast to the marked increase in transportation activity in road freight (see Figure 5.3). A volume of goods that was barely increasing was thus moved farther and even more frequently by road. Increasing integration in the world market has certainly made an essential contribution to a growing proportion of goods being exchanged over ever greater distances. Although this development could have favoured rail transport as the cleaner alternative, this did not occur.

The considerable significance of intrasectoral change (Japan, FRG, Sweden, Portugal)

We have so far been dealing with intersectoral structural change and its implicit environmental impact. In summary, it did not reduce structural environmental stress in the southern and eastern European industrial countries. On the contrary, the industrial environmental situation in these areas deteriorated. In the highly-

National Change Profile Federal Republic of Germany (1960=100)

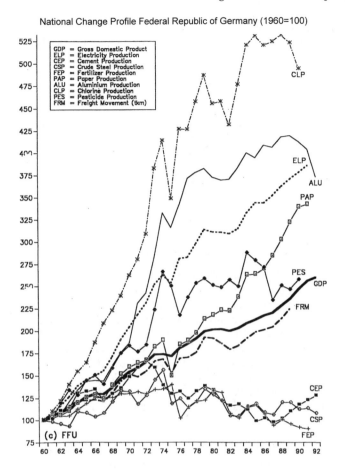

Figure 5.5 National change profile FRG 1960–92.

advanced industrialised countries, too, there was no reduction in environmental stress in the above sense, despite a slight decline in traditional heavy industries (mineral products, crude steel, fertiliser). This was because modern industries with high environmental impact – especially the chemical industry – have experienced strong growth. This is also true for countries with strong structural change such as Japan.

A somewhat different picture emerges if we also consider change within industrial sectors and enterprises. At least in the advanced industrialised countries, this intrasectoral change has hitherto been the most significant factor from the ecological point of view. We have – unfortunately only incomplete – figures for the highly-advanced industrialised countries Japan, the FRG and Sweden. A comparison will be drawn with the less-advanced industrial country Portugal.

With regard to intrasectoral change it can be assumed that technological

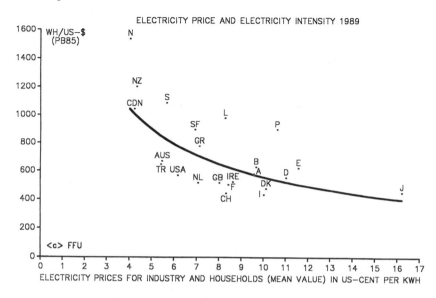

Figure 5.6   Electricity price and electricity intensity 1989.
  Abbreviations: A, Austria; AUS, Australia; B, Belgium; BG, Bulgaria;
  CDN, Canada; CH, Switzerland; CS, Czechoslavakia; D, West Germany;
  DDR, East Germany; DK, Denmark; E, Spain; F, France; GB, Great
  Britain; GR, Greece; H, Hungary; I, Italy; IRE, Ireland; IS, Iceland; J,
  Japan; L, Luxembourg; N, Norway; NL, Netherlands; NZ, New Zealand;
  P, Portugal; PL, Poland; R, Rumania; S, Sweden; SF, Finland; SU, Soviet
  Union; TR, Turkey; USA, United States of America; YU, Yugoslavia.

change predominates. However, there are also non-technological changes in the relative weighting of product groups within industries, which have to be taken into consideration. If, for example, products in the chemical industry that have always been relatively cleaner to produce than others gain in importance, this can no more be considered modernisation. The same is true for the relocation of problematic production processes; we thus refer to intrasectoral change.

We measure intrasectoral change as an adjustment in a number of the above factors – primarily changes in energy and water intensity. Figures are also provided on the usually marked, but differentiated, trend towards electricity use (electricity use is included in data on energy consumption at the same time). Data are also available on raw-material consumption and on waste production by the manufacturing sector in the FRG and Japan. In addition there are industrial land-use figures for Japan. We will proceed to describe intersectoral and intrasectoral change in the four countries mentioned, showing developments in Japan and the FRG in Figures 5.7 and 5.8.

Japan experienced considerable intersectoral change and incomparably greater intrasectoral change. The relative decrease in significance of mineral products, basic metal production and the paper industry and an absolute down-

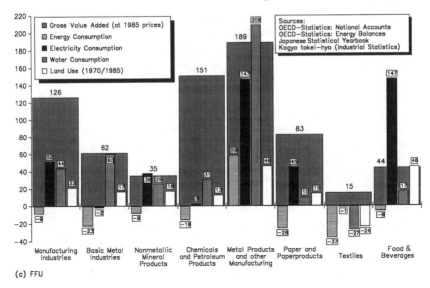

Figure 5.7 Percentage change by indicator and sector 1971–87: Japan.

turn in special basic industries (aluminium, fertiliser) was accompanied by radical change within individual sectors. Although the chemical industry experienced above-average growth, it nevertheless reduced final energy consumption in absolute terms by 16 per cent. Metal production developed similarly. In the textile

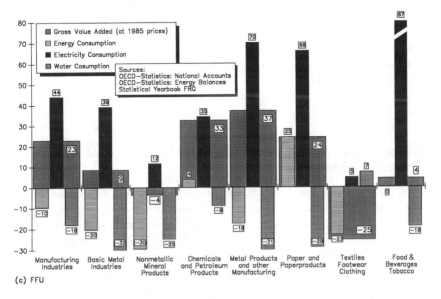

Figure 5.8 Percentage change by indicator and sector 1971–87: FRG.

industry there were improvements throughout. For the 1977–87 period the picture is even more favourable to the environment: stagnation or an absolute drop in energy and water consumption and land use by the metal industry, the chemical industry (with the exception of a 4 per cent increase in land use), the paper industry and the textile industry.

Intrasectoral change in Japan proved to be little less than dramatic: industrial final energy consumption was in 1989 no less than 58.6 per cent lower than it would have been without primarily technological change within sectors after 1970. In the FRG this stress-reducing effect of intrasectoral change made a difference (up to 1989) of 30.4 per cent and in Sweden (1973–88) of 27.6 per cent. In Portugal change within industrial sectors resulted rather in structural deterioration.

By contrast, the reduced significance of energy-intensive sectors (intersectoral structural change) lowered energy consumption in Japan and the FRG after 1970 on balance only marginally. In each case it was by over 13 per cent. In Sweden and Portugal this effect was quite unimportant.

In the case of industrial water consumption, too, technological change within sectors was very much conspicuous. Japan consumed 29.5 per cent less water and the FRG 36.9 per cent less (1971–87) than they would have used without intrasectoral change. In Sweden, where precise sectoral data on water consumption are not available, a marked reduction was also only possible because of conservation within industries. The intersectoral components in Japan, by contrast, produced a difference of only 8.2 per cent (1970–87). In the FRG and Sweden it was of as little significance.

With regard to waste produced by the manufacturing sector, intersectoral change brought a reduction in the FRG. There was a 10.4 per cent reduction in waste over the period 1977–87. Intrasectoral change, however, resulted in a 19.8 per cent deterioration. The growth-related rise in industrial waste production was thus not compensated for, but – also as a consequence of environmental protection measures – aggravated, producing a 23.5 per cent increase. In Japan too, industrial waste increased in absolute terms (32 per cent during 1975–85), but in contrast to the FRG the rise was slower than the parallel increase in industrial value added.

Reduction in environmental stress by means of industrial structural change?

What, then, was the impact of structural change on the four countries we have been looking at in detail in relation to the above-mentioned ecologically significant criteria? The indicators used are industrial energy, water and raw-materials consumption, production of waste and the volume of freight movement by rail and road (the latter is concerned not only, but primarily, with industrial goods).

The greatest reduction was in industrial final energy consumption. The three highly-advanced industrialised countries Japan, Sweden and the FRG experienced an absolute reduction. Industrial water consumption also showed a relatively favourable trend. In the FRG and Sweden there was also a drop in absolute terms. In respect of both indicators there was thus a real reduction in structural environmental stress. In Japan a specific reduction in both fields was much greater. Nevertheless, strong industrial growth led to an absolute increase in water consumption. However, from 1979 onwards consumption stagnated.

The relative or absolute reduction in industrial energy and water consumption contrasts with the absolute deterioration in waste production (Japan, FRG). Nearly the same is true with regard to freight movement (three countries), with the exception of Sweden, where the transport activity was nearly stabilised after 1974. Japan showed at least a drop in intensity for all four indicators, thus achieving an uncoupling from economic growth. This was so substantial that for a long period there was no deterioration in the growth process. This observation also applies with regard to a further indicator, industrial land use: after 1975 it hardly increased at all. Industrial land in Tokyo even decreased in area by 23 per cent between 1975 and 1985 (Tokyo Metropolitan Government 1987: 283).

Data are also available on raw-materials consumption in the FRG, according to which consumption dropped slightly in absolute terms in the 1980s (Statistisches Bundesamt *Statistisches Jahrbuch* 1994: 742ff.; see also Binswanger (1993)). In Japan there was an increase of about 50 per cent during 1970–90 (Environmental Agency 1992: 156); despite this, consumption almost stagnated between 1975 and 1985.

According to the five indicators selected, structural change in advanced industrialised countries has thus followed a very ambivalent course. This also became apparent from the rough classification used for the nine polluting branches of the economy (see above). On balance we cannot speak of reduction in structural environmental stress. However, from the ecological point of view, without structural change things would have deteriorated dramatically. This is probably the most important result to be attributed to industrial change since the oil crisis.

The relative benefits of the alteration in the growth pattern after 1973 only become clear when being tested against the case of Portugal. Even with the unsatisfactory data situation it is obvious that industrial energy and water consumption rose not only in absolute terms but also relative to economic performance. This also applies with regard to freight movement.

On the other hand, changes in Japan, constituting the largest reduction in structural environmental stress worldwide, were strongly counteracted by rapid industrial growth. From 1986 onwards energy consumption (especially in the form of electricity) once again increased conspicuously. The same is true for raw-materials consumption and road and rail freight, which until 1985 had hardly risen above the 1972 level (see Figure 5.9).

The best-practice case of Japan thus demonstrates two things. Firstly, it becomes clear how extensive reductions in important input factors can be effected primarily by means of adjustments within sectors. For a period, Japan was almost a country with qualitative growth sufficiently uncoupled from ecologically relevant input factors. Secondly, however, it became apparent how difficult it is to sustain such stress-reducing uncoupling effects in the long-term growth

GROSS VALUE ADDED AND RESOURCE USE BY MANUFACTURING: JAPAN (1970=100)

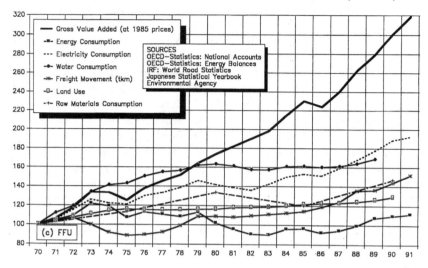

Figure 5.9　Resource use by manufacturing in Japan 1970–91.

process. The formulas of qualitative growth and sustainable development had hitherto hardly been associated with the notion that the reduction of environmental stress in the growth process has constantly to be renewed and extended.

The Portuguese growth pattern, representing a large number of less-advanced industrialised countries but also the growth pattern in Eastern Europe has hitherto experienced, would, if universalised, prove ecologically disastrous (Paulus 1993). But the Japanese growth pattern – with its partial relocation or export of environmental problems – does not yet offer an alternative, environmentally appropriate approach to production processes that are globally generalisable.

The practical absence of ecologically-motivated industrial policy

We also examined the role played by industrial and economic policy in possible ecological structural improvements in the four industrialised countries described (we did not examine the structural policy effects on energy, transport and technology policies).

We found that intersectoral change had been specifically influenced so far only in Japan. It is true that after 1973 energy policy in other industrialised countries also brought about actual reductions in the mineral oil industry. But ecological considerations generally played at best a subordinate role. Moreover, in a considerable number of countries – the US, the UK and Canada, for example – other types of industrial policy are officially strictly rejected.

By contrast, Japan, the country which has experienced the most far-reaching structural change, not only espoused industrial structural policy, but also gave it an ecological bias (Foljanty-Jost 1990). The 1971 Ministry of International Trade and Industry (MITI) developed the concept of a know-how intensive-production structure, with its call for material- and energy-intensive production to be rejected. Although primarily motivated by economic-policy considerations, it had distinct ecological implications – accompanied by forced-pace environmental policy. After the oil crisis, MITI in 1974 called for the traditional growth model based on branches with high material consumption to be abandoned. Increasing attention was paid to the environmental question. Finally, in 1978, legislation provided concrete measures dealing with sectors experiencing economic and ecological problems. Subsequently, the production of primary aluminium was drastically reduced. However, aluminium was then largely imported – with a high recycling rate. Fertiliser production continued to decrease in orderly fashion. Further regulation affected synthetic fibres, petrochemicals, electrosteel, shipbuilding, paper and textiles. These were all sectors with high rates of material, energy and environmental consumption. Overall priority was given to reducing import dependence, but the environmental question was not without significance as a further motive.

At the beginning of the 1970s (even before the oil crisis) there was a certain amount of discussion in the FRG on the relation between industrial structure and pollution. The 1971 Annual Economic Report issued by the federal government stated: 'In its structural policy the federal government will seek improvements in environmental conditions to a greater extent than before' (*Jahreswirtschaftsbericht* 1971: 77). But an about-turn occurred in the period after 1975. Economic policy took the opposite direction under the banner of structural crisis: maintenance subsidies, not ecological concerns, became dominant. In the course of time subsidies even increased. In 1987 they accounted for one third of all federal subsidies (Stille 1990: 15); in 1992 for more than 40 per cent (DIW 1992: 617). In the late 1980s, however, analysis of structural economic concerns espoused the topic of ecology (RWI 1987,1992; Halstrick 1988; Härtel *et al.* 1987).

Sweden, with its pronounced industrial and environmental policy (Jänicke 1992), permitted the assumption that a structural policy motivated by, among other things, ecological considerations would be started. Initially this was indeed the case – at least as far as general statements of intention were concerned. But we find a cycle of discussions on ecological questions strikingly similar to that prevailing in the FRG: in the early 1970s long-term economic planning took into account the relation between pollution and industrial development. As in the FRG, however, discussion was all that happened. After the oil crisis the question dropped out of sight in both countries. And only in recent years has the question of an ecologically more appropriate industrial structure received greater attention in economic structural policy.

The ecological effects of Swedish industrial development have thus been almost as ambivalent as they have been in the FRG. Figures issued by the Swedish Central Office for Statistics show no favourable ecological balance for intersectoral structural change. But here, too, intrasectoral change (not least of all technological change) in polluting sectors such as the paper industry has been of greater significance. Altogether, it was in Sweden that slow industrial growth had better ecological effects than in Japan, although the industrial change was less radical.

The question of ecologically-influenced industrial policy in Portugal is not precluded from the outset, since this country on the southern periphery of Europe established environmental policy institutions at a relatively early date. Protection of the environment was entrenched in the constitution in 1976; and as early as 1974 a Ministry for Social Affairs and the Environment was set up. Ecologically-oriented structural policy was, however, not on the government's agenda. The EC Regional Fund thus tended to have a structurally unfavourable impact (support for the paper industry). It was only in the Basic Environment Law of 1987 that ecological interests in industrial development were given a sort of general-clause mention. Nevertheless, in the 1987 energy plan lower growth rates in energy-intensive sectors were scheduled. In 1988 economic structure reports gave programmatic status to the restructuring of the Portuguese economy in order to conserve resources and energy. Up to this date, however, Portuguese industry did not develop accordingly.

Conclusion

Although the topic of the relation between environmental policy and industrial policy had been taken up in advanced industrialised countries – like Japan, Sweden and the FRG – as early as the beginning of the 1970s, this bore little real fruit except in Japan (Meadows *et al.* 1992). Until well into the late 1980s environmental policy was essentially a special, end-of-pipe, policy. Only in Japan was attention given to the deliberate and socially cushioned reduction of polluting industrial sectors. And there, too – as elsewhere with the mineral oil industry – energy-policy considerations were to the fore.

By contrast, autonomous structural change in the traditional heavy and basic industries in advanced industrialised countries (crude steel, cement, fertiliser) led to reductions in structural environmental stress. But from the point of view of environmental policy these were to be regarded more as chance side effects. And they were cancelled out – just as much by chance – by disproportionately high growth in industries causing high structural environmental stress, such as chlorine, aluminium or paper and paperboard. However, in many places these industries have been under fire from public opinion, which does not preclude a change in patterns of demand behaviour. But this, too, would constitute a form of autonomous structural change.

The greatest structural environmental stress reduction effect in the advanced industrialised countries was caused by (primarily technological) changes taking place within sectors and enterprises (intrasectoral change). But such changes, too, seem for the most part to be autonomous, triggered especially by alterations in price patterns. In this respect Japan is also the only country to show a certain degree of regularity in relative structural environmental stress reduction for parameters such as energy, raw-materials and water consumption, land use and freight movement. However, Japan's high industrial growth rate appears to be the cause of the reductions achieved being cancelled out in the course of time.

Whatever general problems ecologically-motivated industrial structural policy may present, the following conclusions are, in our opinion, justified by the evidence. First, our investigations lend support to demands that energy be made significantly more expensive by imposing energy taxes (or in the case of innovation processes by levying charges to be ploughed back as extensively as possible). This not only brings reductions in ecologically-sensitive energy requirements, it also has structural effects that are to be classified as ecologically beneficial. A concentration of such taxes or charges on electricity seems to be particularly appropriate since there is unexploited conservation potential in this field. Secondly, subsidies that have ecologically detrimental effects need to be rejected. This leads to industrial structural policy becoming taboo. The minimum recommendation would be that governments adhering to this ought also to waive every type of ecologically detrimental subsidy, including, for example, subsidies for steel production or coal mining. Reducing these subsidies would probably be equivalent to an environmental-stress reduction effect in many industrialised countries – not to mention fiscal relief effects.

Thirdly, active promotion of economic growth needs to be questioned. The example of Japan shows that ecologically-beneficial economic change tends to be neutralised by high growth. Growth rates themselves are an environmental problem. It is apparent that qualitative growth can in the long term only be limited growth, if ecologically negative growth effects are to be compensated by technological and structural change. If with an annual growth rate of 1 per cent we have a doubling in 70 years, with 5 per cent growth this factor would be attained in a mere 14 years, and in 70 years it would be thirty-fold. Such growth clearly cannot continue to be compensated for by structural effects; except by relocation of production, which does not provide a global or long-term solution. For this reason the state must rethink the role of its economic policy as the engine of economic growth. The industrialised countries will not be able to afford the luxury of high growth rates for much longer. They will have to become accustomed to solving universal problems not by economic growth, but by political action, as in matters of distribution. The transition from quantity (financial flow) to quality (political structuring) may be also a task for the state.

Notes

1 The research project (1989–94) was sponsored by the Volkswagen Foundation. The follow-up project (1996–2000), sponsored by the European Commission and the Deutsche Forschungsgemeinschaft, focussed on the private and public management of declining dirty industries in six European countries. Binder, M., Jänicke, M. and Petschow U. (eds.) (2000): *Green Industrial Restructuring: Case Studies and Theoretical Interpretations*, Berlin: Springer

References

Ayres, R. U. (1992) *Industrial Metabolism. Theory and Policy*, paper FS II 92–406, Berlin: Wissenschaftszentrum Berlin für Sozialforschung.

Binswanger, M. (1993) *Gibt es eine Entkopplung des Wirtschaftswachstums von Naturverbrauch und Umweltbelastungen? Daten zu ökologischen Auswirkungen wirtschaftlicher Aktivitäten in der Schweiz von 1970 bis 1990*, Institut für Wirtschaft und Ökologie Diskussionsbeiträge 12, St. Gallen: Hochschule St. Gallen.

DIW (1992) DIW-Wochenbericht 46/92, Berlin: Deutsches Institut für Wirtschaftsforschung.

DIW/RWI (Deutsches Institut für Wirtschaftsforschung/Rheinisch-Westfälisches Institut für Wirtschaftsforschung) (1993) *Einfluß umweltbezogener Standortfaktoren auf Investitionsentscheidungen*, Berlin: Erich Schmidt Verlag.

Environmental Agency (1992) *Quality of the Environment in Japan*, Tokyo: Environmental Agency.

Fischer-Kowalski, M. (1991) *Verursacherbezogene Umweltindikatoren*, abstract, Vienna: Ökologie Institut.

Foljanty-Jost, G. (1990) *Industriepolitik in Japan – Ansätze für einen strukturpolitischen Umweltschutz?* FFU-Report 90/6, Berlin: Forschungsstelle für Umweltpolitik der FU Berlin.

Garnreiter, F., Jochem, E. and Schön, M. (1986) 'Produktstrukturwandel und technischer Fortschritt als Bestimmungsgrößen des spezifischen Energieverbrauchs in vier energieintensiven Industriezweigen', *Zeitschrift für Energiewirtschaft* 4, 3: 271–83.

Gerau, J. (1978) 'Zur politischen Ökologie der Industrialisierung des Umweltschutzes', in M. Jänicke (ed.), *Umweltpolitik. Beiträge zur Politologie des Umweltschutzes*, Opladen: Leske & Budrich.

Halstrick, M. (1988) 'Entlastung durch Strukturwandel? Zum Zusammenhang zwischen sektoralem Strukturwandel und Umweltbelastung', *RWI-Mitteilungen* 39, 2: 173–92.

Härtel, H.-H., Matthies, K. and Mously, M. (1987) *Zusammenhang zwischen Strukturwandel und Umwelt*, Hamburg: Verlag Weltarchiv.

Jänicke, M. (1979) *Wie das Industriesystem von seinen Mißständen profitiert*, Opladen: Westdeutscher Verlag.

—— (1984) *Umweltpolitische Prävention als ökologische Modernisierung und Strukturpolitik*, IIUG discussion papers 84–1, Berlin: Wissenschaftszentrum Berlin für Sozialforschung.

—— (1992) 'Conditions for environmental policy success: an international comparison', *The Environmentalist* 12, 1: 47–58.

Jänicke, M., Mönch, H., Binder, M. (1992) *Umweltentlastung durch industriellen Strukturwandel? Eine explorative Studie über 32 Industrieländer (1970–90)*, Berlin: Sigma.

—— (1997) 'Dirty Industries': patterns of change in industrial countries', *Environmental and Resource Economics* 9, 4: 467–491.

Johnson, C. (ed.) (1984) *The Industrial Policy Debate*, San Francisco, CA: Institute for Contemporary Studies.

Low, P. (ed.) (1992) *International Trade and the Environment*, Washington, DC: The World Bank.

Lucas, R. E. B., Wheeler, D. and Hettige, H. (1992) 'Economic development, environmental regulation and the international migration of toxic industrial pollution:

1960–1988', in P. Low (ed.), *International Trade and the Environment,* Washington, DC: The World Bank.

Meadows, D., Meadows, D. and Randers, J. (1992) *Beyond the Limits,* Post Mills, VT: Chelsea Green Publishing.

Paulus, S. (1993) *Umweltpolitik und wirtschaftlicher Strukturwandel in Indien,* Frankfurt am Main: Peter Lang.

RWI (1987) *Analyse der strukturellen Entwicklung der deutschen Wirtschaft (Strukturberichterstattung 1987). Schwerpunktthema: Strukturwandel und Umweltschutz,* Essen: Rheinisch-Westfälisches Institut für Wirtschaftsforschung.

—— (1992) *Umweltschutz, Strukturwandel und Wirtschaftswachstum,* Essen: Rheinisch-Westfälisches Institut für Wirtschaftsforschung.

Schmidt-Bleek, F. (1994) *Wieviel Umwelt braucht der Mensch?* Berlin: Birkhäuser.

Simonis, U. E. (ed.) (1989) *Präventive Umweltpolitik,* Frankfurt am Main, New York: Campus.

Stevens, C. (1993) 'Do environmental policies affect competitiveness?' *OECD Observer* 138: 22–5.

Stille, F. (1990) 'Umweltpolitische Auswirkungen staatlicher Subventionspolitik', in U. Petschow and E. Schmidt (eds.), *Staatliche Politik als Umweltzerstörung. Der Staat in der Umweltverträglichkeitsprüfung,* Schriftenreihe des IÖW 37/90, Berlin.

Tokyo Metropolitan Government (1987) *Second Long-term Plan for the Tokyo Metropolis,* Tokyo: Tokyo Metropolitan Government.

Zimmermann, K., Hartje, V. J. and Ryll, A. (1990) *Ökologische Modernisierung der Produktion. Strukturen und Trends,* Berlin: Sigma.

Statistical sources

Bureau of Statistics, Office of the Prime Minister, Japan, *Statistical Yearbook,* Tokyo, various issues.

DIW (Deutsches Institut für Wirtschaftsforschung) (1991) *Verkehr in Zahlen 1991,* Bonn: Bundesminister für Verkehr.

Environment Agency, Government of Japan, *Quality of the Environment in Japan,* Tokyo, various issues.

FAO (Food and Agriculture Organization), *Fertilizer Yearbook,* Rome: FAO, various issues.

—— *Yearbook of Forest Products,* Rome: FAO, various issues.

IEA (International Energy Agency) (1991) *Energy Prices and Taxes,* 4/91, Paris: IEA.

—— (1993) *Energy Statistics and Balances,* diskette, Paris: IEA.

IRF (International Road Federation), *World Road Statistics,* Geneva: IRF, various issues.

INE (Instituto Nacional de Estatística) (1988) *Estatísticas dos Transportes e Comunicaçoes, 1987,* Lisbon: INE.

MARN/MPAT (Ministério do Ambiente e Recursos Naturais/Ministerio do Plano e da Administraçao do território) (1990) *Relatório do Ambiente e Ordenamento do Território 1990,* Lisbon: MARN/MPAT.

Metallgesellschaft AG, *Metallstatistik,* Frankfurt am Main: Metallgesellschaft AG, various issues.

OECD (Organisation for Economic Co-operation and Development) *Main Economic Indicators,* Paris: OECD, various issues.

—— *National Accounts of OECD Countries*, Paris: OECD, various issues.

Statistisches Bundesamt, *Abfallbeseitigung im Produzierenden Gewerbe und in Krankenhäusern*, (Fachserie 19, Reihe 1.2), Stuttgart, various issues.

—— *Statistisches Jahrbuch für die Bundesrepublik Deutschland*, Stuttgart, various issues.

—— *Volkswirtschaftliche Gesamtrechnungen, Konten und Standardtabellen*, Hauptbericht (Fachserie 18, Reihe 1.3), Stuttgart, various issues.

—— *Wasserversorgung und Abwasserbeseitigung der Industrie*, (Fachserie D, Reihe 5), Stuttgart and Mainz, various issues.

—— *Wasserversorgung und Abwasserbeseitigung im Bergbau und Verarbeitenden Gewerbe und bei Wärmekraftwerken für die öffentliche Versorgung* (Fachserie 19, Reihe 2.2), Stuttgart, various issues.

Statistical Bureau of the United States, *Statistical Abstracts of the United States*, Washington, DC: SB, various issues.

Tsusho Sangyo Daijin, Kanbo Chosa Tokei-bu (Abteilung für Erhebungen und Statistik beim Minister für Internationalen Handel und Industrie), *Kogyo tokei-hyo* (Industriestatistiken), Tokyo, various issues.

UIC (Union Internationale des Chemins de Fer), *International Railway Statistics*, Paris: UIC, various issues.

UN (United Nations) *Demographic Yearbook*, New York: UN, various issues.

—— *Industrial Statistics Yearbook*, New York: UN, various issues.

—— *Monthly Bulletin of Statistics*, New York: UN, various issues.

—— *Statistical Yearbook*, New York: UN, various issues.

—— *Yearbook of World Energy Statistics*, New York: UN, various issues.

UNECE (United Nations Economic Commission for Europe), *Annual Bulletin of Transport Statistics for Europe*, Geneva: UNECE, various issues.

UNCTAD (United Nations Conference on Trade and Development), *Yearbook of International Commodity Statistics*, New York: UNCTAD, various issues.

6 Patterns and correlates of environmental politics in the Western democracies

Detlef Jahn

Introduction

Ecological modernisation is a term that has received an enormous amount of attention in recent years. Much of its attraction lies in the fact that it promises to combine two seemingly contradictory concepts: environmental protection and economic growth. 'Instead of seeing environmental protection as a burden upon the economy the ecological modernist sees it as a potential source for future growth' Weale (1992: 76). But how is this concept compatible with the claims about the limitations of growth? In particular the idea of environmentalism or deep ecology (Cotgrove 1982; Dobson 1990) demands decreasing use of natural resources, which is difficult to combine with extensive growth rates; and in the theoretical debate, the concept of modernisation is at the centre of critiques (Beck 1992). In this light ecological modernisation seems to be a contradiction in terms.

Taking this debate as a starting point, I will focus on two aspects. First, to what extent does environmental protection go hand in hand with a change in expansionist economic policy (hereafter referred to as productionism)? Is there an association between environmental protection and a more careful use of resources among the OECD countries? Only a simultaneous change in both factors would meet the needs of environmentalism. Successful environmental protection combined with a continuation of productionism would clearly favour the established political order and would not imply a change in the direction towards a more ecological society. The second aspect of this chapter investigates some of the factors which influence social development (a) towards more successful environmental protection and (b) towards a detachment from productionism.

Success and failure in environmental progress

What is the state of the environment? This question is of course not easy to answer. The complexity of the issue and the lack of comparable data restricts the analysis (Jahn 1998). A focus on air pollution allows me to concentrate on one of the most important and well-documented environmental problems. This issue has been on the agenda of international organisations since the early 1970s, although it did not gain a high public profile before the late 1980s and was debated as early as the UN Conference on the Environment in Stockholm in 1972. The first World Climate Conference which led to the creation of the World Climate Programme by the World Meteorological Organisation took place in 1979. The issue became more extensively known with the release of the report *Our Common Future* by the World Commission on Environment and Development (the Brundtland Commission). The report's chapter on energy choices addresses global climate change as the first among several environmental risks: 'the serious probability of climate change generated by the "greenhouse effect" of gases emitted to the atmosphere, the most important of which is carbon dioxide (CO_2) produced from the combustion of fossil fuels' (WCED 1987: 172). The objective of the report's suggestions was to manage primary energy consumption by efficient measures, thereby allowing the industrialised countries to stabilise their primary energy consumption.

Because of its importance I would like to concentrate on CO_2 emissions. This analysis focuses on developments in the 1980s, because environmental problems during this decade were more prominent and some political actors began to respond to this challenge. This analysis ends in 1989 because changes in Eastern Europe modified the political agenda of most industrial societies and therefore environmental issues were pushed into the background. It must be stressed that this analysis focuses on *progress*, which can be affected by many factors, such as governmental policy, structural change, or economic development, etc. One should also bear in mind that this analysis reflects on *improvements* in environmental performance, which is clearly different from a straightforward comparison to discover which country is the most environmentally friendly. The basis of comparison lies in actual outcomes, not in the means by which these outcomes – intended or unintended – were pursued. An analysis of more specific policies or policy instruments of individual countries would exceed the scope of this chapter, although it is of course important.

Table 6.1 shows the degree of CO_2 emissions and the changes between 1980 and 1989 in each country.[1] An index which estimates the changes regarding the amount of CO_2 emissions is displayed in column 8.

When CO_2 emission rates are compared the USA is responsible for almost half of the Western world's pollution followed by Japan and Germany, which each emit one fifth of that of the US. By referring the data to emissions per capita, one may conclude that the US and Canada are the clear leaders in CO_2 pollution while Australia and Germany are well above the average of the other highly-industrialised countries. Less-industrialised countries such as Turkey, Portugal and Spain are at the bottom of the list, as are the more industrialised countries Switzerland, France and Italy.

Consideration of the changes in CO_2 pollution over the last decade leads to a distinction between those countries that increased CO_2 emissions and those that decreased it. When viewing the increase rates of Portugal and Turkey, it must be remembered that these countries started from a relatively low level of emissions. Together with New Zealand and Greece they increased CO_2 pollution substantially. Norway, Ireland, Australia and Spain are still performing very badly. Japan

Table 6.1 Changes in CO_2 emissions in twenty-two OECD countries, 1980–9

Rank	Country	Emissions 1980[a]	Emissions 1989[a]	Change %	Emissions per capita 1989	Emissions per capita Index	Index of Change	Total index
1	France	144.0	110.8	−23.06	1963.84	0.33	−1.00	−0.50
2	Sweden	25.0	21.4	−14.40	2500.29	0.42	−0.62	−0.15
3	Belgium	39.0	33.6	−13.85	3371.12	0.57	−0.60	−0.02
4	Denmark	18.0	16.2	−10.00	3151.14	0.54	−0.43	0.08
5	Switzerland	12.0	12.1	0.83	1780.46	0.30	0.01	0.24
6	UK	169.0	164.0	−2.96	2856.60	0.49	−0.13	0.27
7	Austria	17.0	17.0	0.00	2202.64	0.37	0.00	0.28
8	Italy	112.0	117.3	4.73	2034.80	0.35	0.08	0.32
9	Germany[b]	304.5	282.0	−7.39	4459.77	0.76	−0.32	0.33
10	Spain	57.0	63.2	10.88	1622.22	0.28	0.17	0.34
11	Finland	19.0	18.4	−3.16	3690.33	0.63	−0.14	0.37
12	Japan	272.0	288.0	5.88	2331.23	0.40	0.09	0.37
13	Netherlands	53.0	53.5	0.94	3578.36	0.61	0.02	0.47
14	Ireland	7.0	8.1	15.71	2312.30	0.39	0.25	0.48
15	Norway	9.0	10.6	17.78	2499.41	0.42	0.28	0.53
16	Canada	130.0	135.5	4.23	5090.16	0.87	0.07	0.70
17	Australia	66.0	76.3	15.61	4465.91	0.76	0.25	0.76
18	Turkey	28.0	43.8	56.43	775.59	0.13	0.90	0.78
19	USA	1410.0	1480.0	4.96	5884.15	1.00	0.08	0.81
20	Greece	15.0	22.7	51.33	2238.66	0.38	0.82	0.90
21	Portugal	8.0	13.0	62.50	1318.59	0.22	1.00	0.92
22	New Zealand	5.0	7.9	58.00	2337.97	0.40	0.93	1.00

Source: OECD 1991.
Notes: [a] in million tonnes; [b] Germany refers in the whole chapter exclusively to West Germany.

has also increased its CO_2 pollution, whilst the rates of increase of the USA and Canada, although rather small, are substantial because of their high starting points.

On the other hand there are those countries that received a negative value, indicating relative improvements. Modest improvement in CO_2 pollution can be observed in the UK, Finland and Germany. More substantial are those improvements in Denmark, Belgium and Sweden. The outstanding country in decreasing CO_2 emissions, however, is France.

In order to estimate the national level of CO_2 emissions dynamically, I constructed an index that includes the basic amount of CO_2 emissions in 1989 *and* the changes over time.[2] The results of this analysis, shown as total index in Table 6.1, show that France, Sweden, Belgium and Denmark are the most successful countries in combating CO_2 pollution, while New Zealand, Portugal, Greece and the US are the least successful. How is this result connected to the productionist development of Western societies? I will consider this aspect in the following section.

A change of expansionist development?

The dominant social paradigm of industrial society will be called *productionism*, since production is at the heart of these societies. Productionist development is characterised by the desire to improve social welfare by improving material wealth through the exploitation of natural and human resources. Productionist development is concerned with the increase of economic growth rates, the centralisation of production, large-scale production and so forth. In contrast, the new environmental paradigm questions these societal goals and favours oppositional concepts such as decentralisation and small-scale production.[3]

In order to come to an operationalisation of the degree of productionist development,[4] I would like to propose that we identify different patterns of societal development by focusing on *energy policy* (following Lindberg 1977). The results of these studies indicate very little evidence that Western societies would detach themselves from productionist development. The only partial exception was found in Sweden which decreased its growth rates in energy consumption. However, Table 6.2 bears some surprises.

Looking at the total primary energy supply per capita in 1989, it clearly shows that Canada and the US are the leading nations. They use almost ten times more energy per capita than, for instance, Turkey and almost twice as much as Germany. European countries that use more energy than the average are primarily Finland, Sweden, Norway and Belgium. On the other side of the globe, Australia also belongs to the group of countries with high energy supply. Countries using less energy than others are less-industrialised countries such as Turkey, Portugal, Greece, Spain and Ireland. Among the highly-industrialised countries Italy, Austria and Japan have a relatively low energy supply.

The increase or decrease in rates of energy consumption shows quite clearly the productionist orientation of advanced Western societies (Table 6.2). Although

Table 6.2 Energy consumption and productionism in twenty-two OECD countries

Rank	Country	Total energy supply 1989			Changes 1980–9			Total index of productionism
		Per capita	Index	Rank	Percentage	Index	Rank	
1	Denmark	3.49590	0.42	9.0	−8.07	−1.00	1.0	−0.47
2	Netherlands	4.38510	0.52	15.0	−5.30	−0.66	2.0	−0.11
3	Germany	4.38130	0.52	14.0	−1.17	−0.15	4.0	0.31
4	Austria	3.12620	0.37	7.0	2.48	0.05	6.0	0.35
5	Italy	2.66780	0.32	5.0	8.20	0.18	11.0	0.40
6	UK	3.69100	0.44	11.0	3.31	0.07	7.0	0.42
7	Switzerland	3.49650	0.42	10.0	6.55	0.14	10.0	0.45
8	Ireland	2.77920	0.33	6.0	11.41	0.24	14.0	0.47
9	Belgium	4.76240	0.57	16.0	1.49	0.03	5.0	0.49
10	Japan	3.28610	0.39	8.0	11.05	0.24	13.0	0.51
11	Turkey	0.89840	0.11	1.0	25.88	0.55	19.0	0.54
12	France	3.89970	0.47	13.0	10.21	0.22	12.0	0.56
13	Spain	2.22050	0.27	4.0	20.86	0.45	18.0	0.58
14	Australia	5.07690	0.61	17.0	6.00	0.13	9.0	0.60
15	USA	7.81180	0.93	21.0	−1.21	−0.15	3.0	0.64
16	Finland	5.83260	0.70	20.0	11.53	0.25	15.0	0.77
17	Greece	2.19660	0.26	3.0	32.60	0.70	20.0	0.78
18	Norway	5.38280	0.64	18.0	16.80	0.36	17.0	0.82
19	Sweden	5.66130	0.68	19.0	15.39	0.33	16.0	0.82
20	Canada	8.36520	1.00	22.0	4.84	0.10	8.0	0.90
21	Portugal	1.55490	0.19	2.0	46.79	1.00	22.0	0.96
22	New Zealand	3.84800	0.46	12.0	35.92	0.77	21.0	1.00

Note: Measured in tonnes of oil equivalent (ToE).

almost all countries have reduced their energy supply in relation to their gross national product (GNP) – the only exceptions are Turkey, Greece and New Zealand – the absolute energy supply increased enormously. For instance, Portugal increased its energy supply in the last decade by almost 50 per cent, while highly-advanced countries such as Sweden, Norway, Japan, France and Finland only increased their energy supply by more than 10 per cent during the 1980s. However, energy supply dropped in several countries. Above all others, Denmark and the Netherlands required less energy in 1989 than in 1980, and in countries such as the US and West Germany energy supply stagnated.

The indicator[5] for the degree of productionism, which includes energy supply and its changes over time, shows that most countries are clearly still working within the framework of the dominant paradigm. The only exceptions are Denmark and the Netherlands. However, since the index is only suggestive and not absolute, we may also include Germany, Austria, Italy, the UK and Switzerland – the top third of the list – in the category of countries which show a less productionist development. However, in this group there are also important differences among the countries, as we shall see later. In the next section I will analyse the relationship between environment performance and productionism.

Environmental progress and productionism

Figure 6.1 shows environmental progress in relation to the degree of productionism. First, it is important to note that poor environmental progress is associated with a high degree of productionism (Pearson's $r = 0.39$), although this correlation is not significant in statistical terms. However, some countries represent different examples. No country with a low degree of productionism performed negatively in terms of environmental progress. Denmark, Switzerland, Austria and the UK are examples of the type of countries which simultaneously reduced CO_2 emissions and the degree of productionism. Other countries with a low degree of productionism are in the middle field of environmental progress: Germany, the Netherlands and Italy. Conversely, there are countries with a high degree of productionism and negative environmental progress. These are the most highly-advanced countries with a post-industrial structure such as the US, Canada, New Zealand and Norway, on the one hand, and countries such as Greece and Portugal that are approaching the level of an industrial society, on the other.

Of all productionist countries Sweden is the only exception. Although Sweden is one of the most productionist countries, its environmental progress is better than most other nations. Belgium and France may also be considered as productionist countries that perform relatively well in terms of environmental progress.

Productionism is dependent on many variables, as we shall see in greater detail in the rest of this chapter. However, one major important difference should be considered at this point by focusing exclusively on the less productionist nations. Productionism is connected to economic performance. Therefore, two comple-

		DEGREE OF PRODUCTIONISM		
		low	medium	high
E N V I R O N M E N T A L	positive	Denmark Switzerland United Kingdom Austria	Belgium France	Sweden
	middle	Germany Netherlands Italy	Japan Ireland Spain	Finland
P E R F O R M A N C E	negative		Turkey Australia	USA Greece Norway Canada Portugal New Zealand

Figure 6.1 Environmental progress and productionism in twenty-two OECD countries.

tely different avenues may lead to less productionism. On the one hand, countries may suffer under severe economic crises and therefore may have a low degree of productionism. On the other hand, if countries perform relatively well economically, but have a lower degree of productionism, we may conclude that these countries detach themselves from productionism for political reasons.

Italy and the UK represent countries that have suffered a severe economic crisis and therefore have become less productionist. Both Denmark and the Netherlands also suffered economic problems, but their strong changes in productionism may lead to the conclusion that not all these changes are due to economic problems but also caused by political factors. Finally, Switzerland, Austria and Germany perform rather well economically whilst simultaneously detaching themselves from productionism. Table 6.3 summarises the different character of these nations by referring to established indicators of economic performance.

Summarising the findings leads to three kinds of developments among environmentally successful countries. Firstly, environmental success follows from the

Table 6.3 Economic performance of less productionist countries in the 1980s

Country	Degree of productionism	Unemployment rate	Inflation rate	Misery index [a]
Germany	0.31	6.7	2.9	9.60
Austria	0.35	3.3	3.8	7.10
Switzerland	0.45	0.6	3.5	4.10
Denmark	−0.47	8.1	6.5	14.60
Netherlands	−0.11	9.6	2.8	12.40
Italy	0.40	10.0	10.6	20.60
UK	0.42	9.2	7.5	16.70

[a] Sum of unemployment rate and inflation rate.

productionism of industrial society: France and, primarily, Sweden stand out in this group. The second group combines good environmental progress with less productionism; most clearly represented by Denmark, Austria and Switzerland. However, Germany and the Netherlands are also less-productionist countries with relatively good environmental progress. Finally, in the UK and Italy there has been a positive impact on environmental progress as a result of severe economic problems. However, what are the factors that may explain these different outcomes? I will turn to this question in the remainder of this chapter.

Factors influencing environmental and productionist politics

Explanations as to why countries develop differently in terms of environmental progress and productionism arise from various approaches which may be subsumed under the headings of structural, institutional and power or resource-mobilisation theory. While the structural explanation identifies basic social, economic and cultural changes in modern societies as the source of political (re)orientation, institutional theories attribute a determining role to political arrangements and procedures. Furthermore, power and resource-mobilisation theory stresses the significance of political opportunity structures and focuses on variables such as the power relations between classes, interest groups, and parties. Finally, I will look at state policy and its impact on policy outcomes. In the following analysis I will only use bivariate analyses in order to comment on the association between some basic variables (see also Jänicke 1995).

Structural change, environmental progress and productionism

The data set out in Table 6.4 suggests that the impact of structural factors on environmental politics is rather modest. The size of the service sector has almost no impact on environmental progress and productionism. The slightly higher positive correlation between the size of the agricultural sector and a negative correlation between the size of the industrial sector and environmental progress

Table 6.4 Correlations of major variables with environmental progress and degree of productionism

Variables	Environmental progress				Productionism			
	r	P	N	Rank	r	p	N	rank
Agricultural sector[a]	0.43	0.047	22	8	0.16	0.235	22	15
Industrial sector[a]	-0.33	0.137	22	11	-0.13	0.290	22	17
Service sector[a]	-0.29	0.193	22	13	-0.12	0.300	22	18
Population density[b]	-0.31	0.156	22	12	-0.53	0.006	22	3
Public opinion[b]	0.26	0.370	12	14	0.51	0.043	12	4
GDP[a]	-0.55	0.008	22	6	-0.24	0.145	22	9
Inflation rate[a]	0.38	0.077	22	9	0.20	0.183	22	12
Unemployment[a]	0.00	0.989	22	19	-0.19	0.198	22	13
Misery index[a]	0.33	0.123	22	10	0.11	0.313	22	19
Social security[c]	-0.70	0.000	21	1	-0.38	0.045	21	6
Neocorporatism[d]	-0.62	0.127	18	5	0.28	0.131	18	8
Social-democratic governments[e]	-0.07	0.750	22	17	0.29	0.094	22	7
Greens[f]	-0.63	0.002	22	3	-0.44	0.020	22	5
Left-libertarians[g]	-0.53	0.012	22	7	-0.59	0.004	22	1
Anti-nuclear movement[h]	0.45	(Eta)	22		0.50	(Eta)	22	
Environmental research 1980[i]	0.23	0.407	16	15	0.19	0.242	16	14
Environmental research 1990[i]	0.09	0.708	19	16	-0.21	0.190	19	11
Electricity price[j]	-0.06	0.804	22	18	-0.13	0.286	22	16
Nuclear energy all countries[i]	-0.65	0.001	22	2	0.21	0.171	22	10
Only countries with nuclear energy[j]	-0.62	0.023	13	4	0.59	0.034	13	2

Sources: a OECD 1992; b Hofrichter and Reif, 1990 for the EC countries, Bennulf 1991; c Lane et al. 1991; d Crepaz 1992; e Merkel 1992; f Müller-Rommel 1993; g Kitschelt 1990; h Kitschelt 1989; i OECD 1991; Jänicke et al., 1992.

Notes: r = Pearson's r (all correlations use r except anti-nuclear movements which use Eta because of ordinal scaling); P = level of significance; N = number of cases; Productionism: see Note 5.

suggests that it is not the structural change towards post-industrial society which contributes to environmental progress, but rather the degree of modernisation and economic wealth. However, in general, a densely-populated country performs better with regard to environmental pollution than other countries. Population density is one of those factors which most clearly explain the degree of productionism. This finding leads to the conclusion that countries with large territories perceive nature as a less-valuable resource than other countries. The above conclusion suggests that economic factors may have a greater impact than sectoral changes. Above all wealthy countries such as Switzerland, the Scandinavian countries, Japan, Germany, the US and Canada may be able to invest money in order to improve their environment in contrast to poor countries such as Turkey, Portugal and Greece.

The data confirm the assumption that the wealthier a country, measured by GNP per capita, the better the environmental progress. The correlation between inflation and environmental progress is lower but still clear. Unemployment, in contrast, does not seem to have a strong impact on environmental progress.

Besides structural and economic factors, cultural changes may also have a profound impact on outcome. Value changes in particular may have led to a stronger emphasis on environmental issues. Inglehart (1992) postulates a strong correlation between post-materialist value orientation and concern about the environment. But does public opinion correlate with the environmental progress of a country?

According to surveys,[6] the populations of Germany, Denmark, Sweden and the Netherlands have the greatest concern about the environment. It is lowest in France, the UK and above all in Portugal and Ireland. There is a moderate, positive correlation between environmental concern and a higher level of environmental progress. However, public opinion and productionism correlate strongly. This means that public opinion may have an impact on the environmental progress of a country, although this impact may not be as direct as suggested here. A deeper analysis shows that public opinion correlates very strongly with the strength of left-libertarian parties ($r = 0.86$), which in turn have the strongest impact on less-productionist politics (see below). However, before considering resource-mobilisation factors I wish to focus on institutional ones. Most important are neocorporatist arrangements and the development of the welfare state.

Neocorporatism, institutionalism, the welfare state and the environment

In the neocorporatist view, economic growth served other societal goals indirectly; examples would be employment, monetary stability, demand simulation and even social equity (Schmitter 1988: 503). Concerning environmental performance, some authors consider corporatism a crucial variable (Jänicke 1995). They claim that environmental interests are taken into account earlier if there is a corporatist policy style.

Offe stresses the point that the interest representation in corporatist societies is conducted by the most-powerful interest groups settling their conflicts at the expense of social categories (groups and issues) that are poorly organised (Offe 1981: 128–9). While the first statement of Schmitter postulates an effective performance under neocorporatist regimes, which in turn may lead to more efficient environmental progress in countries with a high degree of neocorporatism, Offe argues that corporatist arrangements lead to a marginalisation of interests that are not efficiently represented by organisations. Environmental problems have been perceived as one such marginalised interest.

The data suggest that corporatism has a positive impact on environmental progress, but only a weak one on less-productionist politics.[7] An interpretation of this finding may be that neocorporatism solves environmental problems to a certain degree but not necessarily in a less-productionist way. The higher correlation between neocorporatism and environmental progress than between neocorporatism and less productionism confirms the hypothesis of Offe that neocorporatist arrangements depoliticise societal conflicts, but it also confirms Schmitter's point that neocorporatism leads to an effective (environmental) policy.

Let us turn to the impact of the welfare state. What impact does the welfare state have on environmental progress? Do welfare states take care of the environment in the same manner as they do of social and economic issues?

Relating the strength of the welfare state – as measured by social security expenditure, the indicator most often applied in comparative studies – to environmental progress and productionism leads to very interesting results. There is a clear correlation between a strong welfare state and environmental progress. This is the strongest correlation in this investigation. However, when it comes to productionism the correlation falls dramatically. This indicates that welfare states – similar to states with neocorporatist arrangements – deal with environmental issues without challenging the productionist development of industrial society.[8] Or to phrase it differently: welfare states solve the environmental problems within the logic of the productionist paradigm. However, institutional factors may be highly influenced by power relations. In order to grasp this point, I wish to focus in greater detail on resource mobilisation and power relations.

Resource mobilisation, power relations and productionism

Environmental progress and the degree of productionism may be dependent on political factors such as resource mobilisation and the political opportunity structure (Gamson 1975; Kitschelt 1986; Tarrow 1991). While focusing on these factors, we have to take into account the resource mobilisation of classes and the new politics. The former aspect refers to the power position of the labour movement while the latter focuses on new social movements and green or left-libertarian parties.

Along with Touraine *et al.* one may argue that left parties promote productionist politics, since industrial values are inherent in a socialist ideology (Touraine *et al.* 1987). This hypothesis may be confirmed when left-party governments are not significantly different from conservative governments, which, of course, are also oriented towards the productionist paradigm. On the other hand, one may predict that left parties have an impact on better environmental progress and less productionist politics, because they are most open to the ecological demands of all established political actors. This hypothesis hints at the potential for old left–new left alliances which Offe (1985) considers as a possibility for modern societies.

The results show that social-democratic governments[9] do not make any difference to environmental progress. In the countries with a conservative government we can see positive cases as well as negative ones. Sweden and France are positive examples of left governments. However, countries such as New Zealand, Australia and Greece disprove the hypothesis that social-democratic governments protect the environment better than conservative ones. On the other hand, Switzerland and Japan show that relatively-positive environmental progress can also take place under a conservative government. Yet it is also clear that conservative hegemony may lead to poor environmental progress, as seen in Canada and the US. Concerning productionist politics the picture becomes clearer. Only one social-democratic country, Austria, is detaching itself from productionism. Other social-democratic countries such as Sweden, New Zealand, Australia and Greece are still closely connected to this concept. It is interesting that several countries with a conservative government are less productionist. This is true of the Netherlands, Switzerland and Germany.

The results show that left governments do not support better environmental progress but that they promote stronger productionist policies than other governments. This result clearly confirms Touraine's hypothesis of labour's inability to change politics towards greater openness for ecological concerns as long as it follows its own values. However, there may be changes if labour is challenged by new social movements and their power position requires alliances (Dalton 1992; Jahn 1993a).

In the context of environmental progress and productionism the resource mobilisation of new politics is important. The new social movements which emerged in the 1970s and 1980s were crucial catalysts and an indicator for the resource mobilisation of new politics. In most countries the new social movements grew out of a broad network of environmental, feminist, peace, neighbourhood and student movements. These movements aim to change the agenda of industrial society but this is not their only goal. They also aim to move society away from economic growth, centralisation and bureaucratisation, and to achieve a more egalitarian and emancipated society (Dalton and Kuechler 1990). The anti-nuclear-power movements are amongst the most important new social movements in modern society where the conflict between productionism and the new environmental paradigm has become manifest. According to Kitschelt (1989: 25–7), seven countries in this study experienced intense conflict over nuclear energy: Sweden, the US, Denmark, Germany, the Netherlands, Switzerland and Austria.

Anti-nuclear protest has had an important impact on both variables, yet to a very different degree. The countries with a record of protest perform somewhat better in the sphere of environmental issues. France, Japan and Belgium are countries without conflict over nuclear energy while performing well on environment outcomes. In terms of productionism the correlation increases. Only Sweden and the US are countries with a strong degree of conflict but which also strongly follow productionist policies. Denmark, Germany, the Netherlands, Switzerland and Austria have had intense conflicts concerning nuclear energy and are less productionist. Italy and the UK are two countries that are less productionist and without intense conflicts over nuclear energy. This result confirms the hypothesis that Italy and the UK have moved away from productionist policies not for political reasons, but rather as a consequence of economic crisis.

The extraParliamentary protest of the new social movements has also been a resource for environmental parties and small parties on the left. Most green parties originated in new social movements and left-party activists participated in anti-nuclear power demonstrations. What impact do these parties have on environmental progress and the paradigm shifts of the most-advanced industrialised countries?

Populations in countries with a strong green party have a significantly higher environmental concern than in other countries $(r = 0.58)$. As our data of twenty-two highly-developed industrialised countries dictate, there is a clear correlation between green parties' votes in the 1980s[10] and the environmental progress of a country. Turning to the detachment from productionism, we can identify a diminishing impact of green parties. Instead, left-libertarian parties have a higher impact when it comes to modifying the path of productionism. Kitschelt (1989: 9) has coined this term for those parties that 'oppose the priority given to economic growth in public policy making, an overly bureaucratized welfare state, and restrictions placed on participation which confine policy making to elites of well-organized interest groups and parties'. Left-libertarian parties are green parties in some countries such as Austria, Belgium, Germany. But they can also be left parties that opened their programmes to environmentalism and decentralisation, such as the Danish and Norwegian Socialist People's Party or the Swedish Left Party.[11]

The results are striking. The emergence of left-libertarian parties is the strongest single factor explaining the detachment from productionist politics. Their impact is also high on environmental progress (seventh highest impact). Therefore, we may conclude that green parties and above all left-libertarian parties are important forces for change in highly-advanced industrialised countries.

The results of this section can be summarised as follows: in contrast to the power of labour parties, the resource mobilisation of new politics is an important variable for environmental progress and, above all, for a change in productionist politics. While green parties have a strong impact on better environmental progress, the emergence of strong new social movements and left-libertarian parties correlates with a lower degree of productionism.

State policy and the two paths of environmental politics

State policy may have a direct effect on the pattern of ecological modernisation. Three aspects will be considered here: firstly, the correlation between public expenditure on research and development for environmental protection as a percentage of the total research and development budget of a country with environmental progress is weak, although it is somewhat higher on reduced productionism. Secondly, Jänicke et al. (1992: 32–7, 143) attribute a high status for environmental improvements to the prices of electricity. The comparison of our OECD countries shows a moderate correlation between electricity prices and environmental progress and a weak one concerning less productionism.

One crucial variable may be nuclear energy. As we have seen above, nuclear power programmes gave rise to the establishment of anti-nuclear-power protest that in turn has led to green and left-libertarian parties which have had a high impact on changes in policy styles. How does the nuclear-power policy connect to environmental progress and productionism?

Nuclear energy has very interesting correlations with our two dependent variables revealing the different environmental policy styles: one the one hand, countries using nuclear energy perform significantly better in respect to the environment. This may be a result of lower CO_2 emissions from nuclear energy plants. On the other hand, these countries are strongly oriented towards productionism. This positive correlation becomes strikingly clear when only considering those countries that use nuclear energy. This correlation is, together with left-libertarian parties, the strongest single factor explaining productionism.[12] The results confirm strikingly that nuclear energy is a crucial factor for the analysis of the development of highly-industrialised countries in terms of productionism. The two paths become particularly obvious: on the one hand, there are countries with better environmental progress that adhere to productionism, symbolised by increasing energy supply and the extensive application of nuclear energy. These countries are above all Sweden, France and Belgium. On the other hand, there are countries detaching from productionism and performing relatively well in terms of environmental progress, such as Denmark, Austria, the Netherlands and Germany.[13]

Conclusion

Ecological modernisation is an ambiguous term since it aims to combine two contradictory aspects: environmental protection and economic growth. The analysis of twenty-two OECD countries, however, shows different patterns of development. Leaving the unsuccessful environmental countries aside, three avenues can be distinguished. First, there are less wealthy countries with economic problems, without an intense conflict over nuclear energy, no relevant green nor left-libertarian parties and which are less productionist than the average. Furthermore these countries have weak neocorporatist arrangements and a relatively small welfare state. This characteristic conforms to the cases of Italy and the UK. The detachment from productionist politics is based on economic problems and has little to do with a political reorientation and ecological modernisation.

The two other avenues of development by wealthy countries emphasise different aspects of ecological modernisation. On the one hand, there are countries that are environmentally successful but stick firmly to productionism. Increasing energy requirements and a substantial nuclear industry are the characteristics of these countries. France, Belgium and Sweden are the most visible examples. On the other hand, some countries stress ecological aspects more than productionism. These countries limit their energy requirement and the amount of nuclear energy. In these countries nuclear energy is a highly-contested issue and left-libertarian parties receive a substantial amount of votes. All these countries have a high degree of neocorporatist arrangements. Countries supporting this statement are: Denmark, the Netherlands, Germany, Austria and Switzerland. Sweden has all the characteristics without detaching from productionism and is therefore a most interesting 'deviant case' needing further research.

Notes

1 Iceland and Luxembourg have been excluded from this analysis, because of the lack of data or substantial structural changes due to their relatively small size.
2 This has been achieved in the same way as for the degree of productionism (see below). Decrease rates are overestimated by 2.7 in comparison to increase rates.
3 For this dichotomy of paradigms, see, for instance, Dunlap and Liere (1978); Cotgrove (1982); and Milbrath (1993).
4 It is hard to place whole societies either into the productionist paradigm or into the environmentalist one. Furthermore, there is no doubt that all advanced industrial societies are clearly affiliated with productionism. This argument is, therefore, on a much more modest level. Has there been some detachment from the productionist orientation in some societies?
5 The index for productionism is standardised in the way that energy supply scores between 0 and 100. The same has been done for the increase and decrease rates which score between 0 and 100, and 0 and -100 respectively. Since the range between the rate of increase and the rate of decrease differs, the rate of decrease is weighted by 5.8. As regards content, this rating results in an overestimation of countries that decreased their energy consumption. This is analytically justifiable since decrease rates are still the exception and should therefore be stressed. In formal terms the index is calculated as follows:

$$DPP = \frac{(TPES\text{-index} + TPES\text{ change index})}{(Max.TPES\text{-index} + TPES\text{ change index})}$$

where TPES-index = TPES (1989)/Max. TPES.
    TPES change index in case of energy decrease = PDR/Max. PDR and in case of energy increase = PIR/Max. PIR. DPP = Degree of productionism; TPES = Total primary energy supply; Max. TPES = Maximal TPES, i.e. the value for Canada; PDR = Percentage of energy decrease rate; Max. PDR = Maximal PDR; PIR = Percentage of energy increase rate; Max. PIR = Maximal PIR.
6 It is difficult to find reliable data from surveys that are comparable across cultures. Asking the question about the importance of environmental problems is highly biased. The most reliable data source can be taken from a study based on Euro-

barometer surveys (Hofrichter and Reif 1990). I summarised the ranking of environmental issues in the context of eight international and four nation specific issues at three time points (1988 and twice in 1989). In addition we can include Sweden where similar studies have been analysed by Bennulf 1991.

7   The data for the degree of corporatism have been taken from Lijphart and Crepaz (1991) who base their index on the comparison of twelve judgements made by experts.

8   Of course, both variables – neocorporatism and the degree of the welfare state – correlate quite strongly: $r = 0.46$.

9   For the analysis I rely on an index of social-democratic government power applied by Merkel (1992). He attributed an annual score for social-democratic government power. Five points: exclusively social-democratic governments with parliamentary majority; four points: exclusively social-democratic governments without parliamentary majorities (minority cabinets); three points: social democrats as the dominant partner in a governing coalition; two points: social democrats as equal partner in a grand coalition; one point: social democrats as junior partner in a governing coalition; no points: social democrats in opposition. In the same manner I added the data for the US, Canada, Japan, New Zealand and Australia.

10   For the operationalisation, see Müller-Rommel 1993: 98–9.

11   Kitschelt (1989: 11) has not been consistent in his classification of left-libertarian parties. This concerns in particular the Swedish case where he sometimes classifies the Centre Party as left-libertarian, sometimes not. It has been shown, that it is inappropriate to classify the Centre Party as left-libertarian; see Jahn 1993b: 190. Therefore, I used the data from Kitschelt (1990: 183) in which he places the Centre Party outside this classification.

12   The causal relationship is ambiguous since it is as plausible to argue that productionist countries use nuclear energy, as to show that nuclear energy leads to productionism.

13   Switzerland has been excluded here since it has a rather high amount of nuclear energy although its scores are otherwise less productionist.

References

Beck, U. (1992) *Risk Society: Towards a New Modernity*, London: Sage.

Bennulf, M. (1991) 'Väjarnas viktigaste frågor 1968–1990', in S. Holmberg and L.Weibull (eds.), *Politiska Opinioner*, Göteborg: University of Göteborg.

Cotgrove, S. S. (1982) *Catastrophe or Cornucopia: the Environment, Politics and the Future*, Chichester: Wiley.

Crepaz, M. M. L. (1992) 'Corporatism in decline? An empirical analysis of the impact of corporatism on macroeconomic performance and industrial disputes in eighteen industrialized democracies', *Comparative Political Studies* 25, 2: 139–68.

Dalton, R. J. (1992) 'Alliance patterns of the European environmental movement', in W. Rüdig (ed.), *Green Politics Two*, Edinburgh: Edinburgh University Press.

Dalton, R. J. and Kuechler, M. (eds.) (1990) *Challenging the Political Order*, Oxford: Oxford University Press.

Dobson, A. (1990) *Green Political Thought*, London: Unwin Hyman.

Dunlap, R. and van Liere, K. (1978) 'The new environmental paradigm', *Journal of Environmental Education* 9, 1: 10–19.

Gamson, W. A. (1975) *The Strategy of Social Protest*, Homewood: Dorsey.

Hofrichter, J. and Reif, K.-H. (1990) 'Evolution of environmental attitudes in the European Community', *Scandinavian Political Studies* 13, 2: 119–46.

Inglehart, R. (1992) 'Public support for environmental protection: objective problems and subjective values', Paper presented at the symposium on 'The Human Dimension of Global Environmental Change', Chicago, IL, 3–6 September.

Jahn, D. (1993a) *New Politics in Trade Unions*, Aldershot: Dartmouth.

—— (1993b) 'The rise and decline of new politics and the greens in Sweden and Germany: resource dependence and new social cleavages', *European Journal of Political Research* 24, 2: 177–94.

—— (1998), 'Environmental performance and policy regimes: explaining variations between 18 OECD countries', *Policy Sciences*, 31, 2: 107–131.

Jänicke, M. (1995) 'Framework conditions for environmental policy success: an international comparison', in A. Carius, L. Höttler and H. Mercker (eds.), *Environmental Management in Kenya, Tanzania, Uganda and Zimbabwe*, Berlin: German Foundation for International Development.

Jänicke, M., Mönch, H. and Binder, M. (1992) *Umweltentlastung durch industriellen Strukturwandel? Eine explorative Studie über 32 Industrieländer (1970–1990)*, Berlin: Edition Sigma.

Kitschelt, H. (1986) 'Political opportunity structures and political protest: anti-nuclear movements in four democracies', *British Journal of Political Science* 16, January: 57–85.

—— (1989) *The Logics of Party Formation. Ecological Politics in Belgium and West Germany*, Ithaca, NY: Cornell University Press.

—— (1990) 'New social movements and the decline of party organizations', in R. J. Dalton and M. Kuechler (eds.), *Challenging the Political Order*, Oxford: Polity Press.

Lane, J.-E., McKay, D. and Newton, K. (1991) *Political Data Handbook OECD Countries*, Oxford: Oxford University Press.

Lijphart A. and Crepaz, M. M. L. (1991) 'Corporatism and consensus democracy in eighteen countries: conceptual and empirical linkages', *British Journal of Political Science* 21, 2: 235–56.

Lindberg, L. (1977) 'Energy policy and the politics of economic development', *Comparative Political Studies* 10, 3: 355–82.

Merkel, W. (1992) 'After the golden age: is social democracy doomed to decline?', in C. Lemke and G. Marks (eds.), *The Crisis of Socialism in Europe*, Durham, NC: Duke University Press.

Milbrath, L. W. (1993) 'The world is relearning its story about how the world works', in S. Kamieniecki (ed.), *Environmental Politics in the International Arena. Movements, Parties, Organizations, and Policy*, Albany, NY: State University of New York Press.

Müller-Rommel, F. (1993) *Grüne in Westeuropa. Entwicklungsphasen und Erfolgsbedingungen*, Opladen: Westdeutscher Verlag.

OECD (1991) *OECD Environmental Data Compendium*, Paris: Organisation for Economic Co-operation and Development.

—— (1992) *Historical Statistics 1960–1990*, Paris: Organisation for Economic Co-operation and Development.

Offe, C. (1981) 'The attribution of public status to interest groups', in S. D. Berger (ed.), *Organizing Interests in West Europe*, Cambridge: Cambridge University Press.

—— (1985) 'New social movements: challenging the boundaries of institutional politics', *Social Research* 52, 4: 817–68.

Schmitter, P. C. (1988) 'Five reflections on the welfare state', *Politics and Society* 16, 4: 503–15.

Tarrow, S. (1991) 'Struggle, politics, and reform: collective action, social movements, and cycles of protest', Occasional paper 21, 2nd edition, Center for International Studies, Cornell University.

Touraine, A., Wieviorka, M. and Dubet, F. (1987) *The Workers' Movement*, Cambridge: Cambridge University Press.

Weale, A. (1992) *The New Politics of Pollution*, Manchester: Manchester University Press.

WCED (World Commission on Environment and Development) (1987) *Our Common Future*, Oxford: Oxford University Press.

7 The conservative government and sustainable development in the UK: 1988–97

Neil Carter and Philip Lowe

The Conservative government was formally committed to the principle of sustainable development from June 1988 when Mrs Thatcher endorsed the concept in the Toronto G7 summit's response to the Brundtland Report. Subsequently, in a series of documents, the Conservative government attempted to elaborate a strategy for implementing sustainable development. However, few major policy changes intended to benefit the environment were introduced in the first half of the 1990s and the government continued to promote many policies that directly damaged it, most notably the almost unquestioned pursuit of undifferentiated economic growth. Of course, it may be that substantive changes in environmental policy may occur only after a reform of the processes by which policy is formulated and implemented. Even the weakest definition of sustainable development would anticipate that major reforms to the machinery of government will be required if greater integration of environmental and economic policy is to materialise. So, evidence of institutional and administrative reform may be an indicator of future substantive policy change. This chapter examines the extent to which, under the Conservatives, the rhetoric of sustainable development penetrated different levels of government, evaluates their machinery of government reforms and identifies various constraints on the enthusiastic acceptance of sustainability within the British policy elite.

A national strategy for sustainable development

Despite the reaffirmation by Prime Minister John Major of his predecessor's support for sustainable development – most notably by his attendance at the Rio summit in 1992 – official involvement in the domestic debate was rather low key. The White Paper on the Environment, *This Common Inheritance* (DoE 1990), and *Sustainable Development: the UK Strategy* (HM Government 1994) were the key documents in the tentative elaboration of the implementation strategy of the Conservative Government. Beyond central government, at the level of quangos and local government, there was generally a more enthusiastic response although the quality of the debate has been patchy.

Following Brundtland, the government commissioned the environmental economist, David Pearce, to advise on its implementation – a process that resulted in the so-called 'Pearce report' (Pearce *et al.* 1989). Although the government made an initial, rather crude, stab at fleshing out the concept in its 1989 document *Sustaining Our Common Future* (DoE 1989), real progress was delayed until the party politicisation of the environment, during 1988–9, pressured the government into producing *This Common Inheritance* in September 1990. The White Paper set out the general principles that should guide future policy: measures should be based on best scientific and economic evidence; precautionary action should be taken where justified; many problems require international solutions; and the public needs greater access to environmental information. Nevertheless, this substantial document was coolly received by environmentalists and the media due to the paucity of new policies, commitments, targets or even a coherent strategy. It was apparent that many of the ambitious and radical proposals contemplated by the Secretary of State for the Environment, Chris Patten, during the process that produced the White Paper, ran aground on the sands of Whitehall departmentalism (Rose 1991). Powerful producer interest groups – farmers, industrialists, road builders, car manufacturers, energy utilities – were ably represented by their 'sponsoring' departments in Whitehall. Despite being denuded of radicalism, the White Paper nevertheless was a milestone, being the first comprehensive statement of Britain's environment policy in one document, even though it was largely a compilation of existing measures, policies and commitments (Lowe and Flynn 1989).

The second important development was the publication of *Sustainable Development: the UK Strategy* in January 1994, the culmination of UK involvement in the Rio 'Earth summit' 2 years earlier. Following the summit, EU Member States made a commitment in Lisbon in June 1992 to prepare national action plans for the implementation of Agenda 21 – the key document agreed at Rio – by the end of 1993. Agenda 21 stated that:

> Government [...] should adopt a national strategy for sustainable development based on [...] the implementation of decisions taken at the conference, particularly in respect of Agenda 21 [...] its goals should be socially responsible economic development while protecting the resource base and the environment for future generations.
>
> (The Earth Summit 1993: 8.7)

A similar commitment was made by the G7 countries at the Munich summit in July 1992.

The Department of the Environment (DoE) launched a year-long consultation process with a seminar in Oxford in March 1993 at which an outline strategy was discussed. This meeting was followed by the publication of a consultation document in July 1993, inviting responses by the end of September from a wide range of organisations on what form the UK's national sustainability strategy should take. In addition, a series of meetings were held by government departments to discuss outline chapters for the final report. Superficially these procedures conformed to the emphasis that Agenda 21 placed on the role of public participation and consultation in elaborating national strategies for sustainability. However, the UK consultation process was rather hurried – little more than 2 months was allowed for the submission of responses – and, in the judgement of many participants, had minimal influence on the final report. By common consent the consultation paper was remarkably bland: it simply outlined in very broad terms the problems needing to be addressed, without offering any new proposals to provoke responses from interested parties. Indeed, the paper did not even set out a draft official position on sustainability. Instead, views were invited on 'the key concepts and principles inherent in sustainable development that should inform and guide decision makers in the public and private sectors and members of the public' (DoE 1993a: 11). This request to identify first principles might be thought strange coming from a government that had been at least nominally committed to sustainable development for several years and had elaborated its own understanding of the concept in several documents. With the consultation process clearly being given a low priority by most departments, many environmentalists feared that the final report would be equally bland and non-committal.

In the event, the *Strategy* largely fulfilled the gloomy expectations of the environmental lobby. It contained no significant new policies; existing commitments were simply repeated. There was no plan with targets and monitoring procedures along the lines of the Dutch environmental plan. Although the document adopted a long, 20-year time horizon – necessary to transcend short-term political considerations – it contained no explicit follow-up mechanism to allow the strategy to be updated. Yet an updating process is particularly vital in view of the remarkable number of loose ends in the document: the calls for 'more research', the promises of 'further guidance' and the need for detailed policies to address specific problems (Young 1994). This vagueness seemed to permeate the government's overall understanding of sustainable development: the *Strategy* provided no specific long-term aims to show what sustainable development would look like in individual policy areas. The absence of this kind of clear vision raises doubts about whether it can really be called a strategy at all.

The weaknesses that pervade the *Strategy* are illustrated by the chapter on agriculture (Whitby and Ward 1994). Sustainability implies that agriculture should be regarded not just as a form of primary production operating within environmental objectives, but equally as environmental management in its own right. However, rather than seek full integration of production and environmental objectives, the *Strategy* treats them as separate spheres whose conflicting demands need to be balanced. On the production side, considerable faith is placed in a direct relationship between the level of farm price support and the environmental pressures from agriculture, thereby allowing the government to present its efforts to reduce the levels of Common Agricultural Policy (CAP) support as the central plank of its sustainability strategy. But the relationship between price, trade reforms and sustainability is complex and uncertain: it may (or may not) lead to a deintensification of production, which in certain circumstances may be environmentally beneficial. On the other hand, in the *Strategy*, environmental problems arising from agricultural practices – such as water pollution or loss of wildlife – stand for themselves: solutions are left to environmental (not agricultural) policy, such as the use of end-of-pipe technologies or the establishment of special protection areas. Environmental protection is still regarded as no more than a luxury add-on, affordable only once an 'advanced development of [...] agriculture and general economic prosperity' (HM Government, 1994: 15.3) has taken place. The smattering of environment-friendly farming practices outlined in the document – such as Environmentally Sensitive Areas and the proposed Habitat Scheme – are voluntary and depend on financial incentives. Taken together they cover only a fraction of farmland. The *Strategy* does not even address the institutional structures and mechanisms needed to direct the agricultural sector along a more sustainable path.

Overall, the environmental lobby was angered and frustrated that the consultation process did not appear to have influenced the final document markedly. Yet, ironically, in the spirit of Agenda 21, three new initiatives to encourage further consultation and dialogue were outlined: a Government Panel on Sustainable Development, a UK Round Table on Sustainable Development and a Citizens' Environment Initiative. Much needed internal reforms of Whitehall were eschewed in favour of external embellishments to the machinery of government. No wonder the general response of environmental groups was one of cynicism.

By the mid-1990s, therefore, the UK government had formally in place a national strategy for achieving sustainability. But it remained an open question to what extent official commitment to the principle of sustainable development heralded any real change in policies or to the processes of government decision-making. In the following sections, various forms of evidence are examined in order to judge how far the concept of sustainable development had penetrated the policy process under the Conservative government.

Learning the rhetoric

Growth [...] the happiest word in the English language.
John Major (*The Independent*, 4 March 1993)

Within government both politicians and bureaucrats have been remarkably slow in learning the new rhetoric of sustainable development. It is not surprising that politicians have rarely modified their language to a more 'politically correct' ecological discourse. John Major's eulogising of the concept of growth is a particularly gruesome example of the low priority given to sustainability (ironically, he also frequently boasted that he was the first head of state to promise to attend the Rio summit), but it helps explain the (lack of) thinking behind the litany of government policies that 'reflected short-term political and economic pressures rather than the longer-term strategies required for sustainable development' (Blowers 1993: 778). Less forgivable is the impermeability of the Whitehall policy-making process (with the exception of parts of the DoE) to the vocabulary of sustainability. Until the Environment Act 1995, the principle of sustainable development had been formally incorporated into just one piece of legislation, the Natural Heritage Scotland Act 1991, which established Scottish Natural Heritage. Even here the government refused to include a statutory definition of the term, thereby leaving it to the discretion of the new body to interpret the practical implications of the principle.

Away from the centre, some quangos sought to turn theory into practice. For example, there was a vigorous debate about the issue of sustainable tourism and the English Tourist Board (1992) actively encouraged all tourist boards to develop environmental action plans; English Nature (1993) produced a sophisticated position statement outlining its notion of environmental sustainability, which is a working concept geared to the maintenance of biodiversity; within the research councils funding academic work, major research programmes were launched on sustainable transport, clean technology, global environmental change and sustainable agriculture.

It is at the level of local government that debate on implementing sustainable development proved most vigorous. In part, this can be explained by the rather ambivalent 'hands-off' approach of central government towards local government. When the White Paper became a victim of Whitehall departmentalism in 1990, it produced the paradoxical result that while the Secretary of State for the Environment was unable to deliver an ambitious pan-governmental strategy, it did allow him to channel his radicalism into issues that were entirely within the remit of his own department. This included both local government and the land-use planning system for which the DoE is responsible. The White Paper raised some interesting issues about the way in which statutory planning could contribute to sustainable development, for example by reducing greenhouse gases. Although the government then declined to include the principle of sustainable development in the Planning and Compensation Act 1991, this important piece of legislation should still, in practice, enhance it. For the Act requires that all district councils produce local plans for their whole areas, which should then be the main determinant of planning decisions. Subsequently, and continuing this uncharacteristic central government enthusiasm for planning, a Planning Policy Guidance Note from the DoE listed among the requirements to be included in each local plan: policies for the achievement of sustainable development; evidence that an environmental appraisal of development-plan options has been undertaken and that environmental considerations have been taken account of in drawing up all policies; policies demonstrating the increased weight to be attached to environmental considerations; and policies for applying the precautionary principle to environmental impacts which may be difficult to undo (DoE 1992). Local authorities, at least in their planning functions, thus received considerable encouragement to pursue sustainable development themes from central government.

In other respects, however, the Conservative government failed to provide leadership for local government. Numerous authorities introduced environmental audits, green strategies and local environmental charters quite independently of central government initiatives. Local authorities were contributing to the debate – and producing charters – well before the publication of the White Paper (Ward 1993). Indeed, many were keen to go further, but were handicapped by the lack of resources and the absence of a clearly defined role for them in the White Paper. Consequently, local authorities increasingly looked elsewhere for leadership and partnership (Ward and Lowe 1994).

One source was the movement towards greater European integration. The Environment Directorate in Brussels, at least when Carlo Ripa de Meana was the Commissioner, was looking to build constituencies for the development of a more ambitious EC environmental policy. A crucial element was the publication in 1990 of the EC's Green Paper on the Urban Environment in which the Environment Directorate made a conscious bid to move into the field of urban planning and to make links with local authorities across Europe. This thinking was taken forward in the EC's Fifth Environmental Action Programme, *Toward Sustainability*, passed in December 1992, to guide EC policy for the rest of the decade. In this programme the Commission hoped to redirect the debate around the issue of subsidiarity to what it described as the 'principle of shared responsibility', i.e. the appropriate degree and form of co-operation between the Community, member states, local authorities, enterprises and citizens' groups in resolving particular environmental problems. This concept of 'administrative subsidiarity' is obviously attractive to local authorities. Indeed, it has been emphasised that 'local authorities are responsible for implementing 40 per cent of the fifth action programme' (Local Government Management Board, 1993: 28). A second catalyst for this debate was Agenda 21 with its specific emphasis on the role of local government in sustainable development (Stewart and Hams 1993).

Certain environmental issues have always been local – notably planning and amenities – but the idea that local government can contribute to resolving global problems has proven very exciting to many local councillors and officials. After many years of cuts in funding, centralisation and diminution of powers, local authorities see sustainable development as offering opportunities to develop a new, positive role.

Greening central government

One initiative in the spirit of sustainable development that emerged from *This Common Inheritance*, and which was further developed in the *Strategy*, involved a series of machinery-of-government reforms aimed at integrating environmental considerations into decision-making across all policy areas: two environmental cabinet committees were formed; 'green ministers' in every department were identified as responsible for environmental matters; annual departmental reports would establish progress in following up the White Paper; the DoE would publish detailed follow-up reports on the White Paper and guidance on how departments should evaluate the environmental implications of policies was to be published. Reform of the machinery of government is always a slow moving process but the early prognosis for these initiatives is poor.

One problem in evaluating the effectiveness of the White Paper reforms is their lack of transparency. It was reported that the two new cabinet committees met just once and twice, respectively, in the first year of their existence (*The Observer*, 11 September 1991). But the number of full committee meetings may be a poor indicator of effectiveness. Government officials pointed out that cabinet committees only meet when there is a serious dispute between departments. Thus, the committee seems to have been 'little more than a papers circulation list' during the process of drawing up the strategy document, with ministers called in only to resolve disputes on agricultural and transport issues (Young 1994). Normally, as an Environment Minister has noted, work is undertaken in bilateral negotiations: 'hardly a week goes by without a meeting, but each meeting will not include everybody' (House of Lords, 1995: 55). Moreover, the presence of a committee at the highest level of government suggests that the environment was not being completely ignored, that an institutional mechanism existed for departments to put issues of concern on the government's agenda, and that all ministers were aware that they may have had to account for the environmental impact of their policies. However, it has been suggested that if the DoE was promulgating an active, radical agenda then there would be plenty of head-to-head confrontations between government departments that a cabinet committee would be required to arbitrate (Hill and Jordan 1994). Nor did the committee produce integration and co-ordination. For example, the day after the publication of the consultation paper on sustainable development in July 1993, the Secretary of State for Transport announced plans to widen the M25 orbital road around London to fourteen lanes in some sections, a proposal that formed only part of a massive £24 billion, 10-year programme of road building to help cope with the anticipated doubling of road traffic in the following 35 years.

Whilst the effectiveness of the cabinet committee is open to debate (if only because of the lack of evidence), the green ministers innovation seems to have contributed almost nothing to improving policy integration. Perhaps the fact that it took as long as ten weeks to put together the 'green team' should have rung alarm bells; subsequently it met no more than three or four times per year (House of Lords, 1995: 55). More fundamentally, there seems to have been confusion at the heart of government concerning the role of green ministers. The strategy document declares that the job of the individual green minister is 'to ensure that environmental considerations are integrated into the strategy and policies of that Department' (HM Government, 1994: 197). Yet subsequently, in evidence to the House of Lords Environment Committee, John Gummer conceded that green ministers did not hold policy meetings, but were, instead, concerned primarily with green 'housekeeping' issues, such as improving energy efficiency and encouraging 'green procurement' within their own departments (House of Lords, 1995: 55). If this latter definition holds, then the green minister initiative can be seen to have played only the most minimal role in co-ordinating environmental concerns across government.

After 1990 there was a more or less annual follow-up report reviewing progress on implementing policies outlined in the White Paper and setting new targets – a practice unique to environment policy. Early evaluations of the newly introduced departmental annual reports found that 'Departments have largely failed to reflect their white paper commitments, and have failed to show integration of environmental concerns with their Departmental objectives and activities' (Green Alliance, 1991); a conclusion reaffirmed a year later (Green Alliance, 1992). Lastly, although the document *Policy Appraisal and the Environment* was published in 1991, little evidence exists that it was being used, particularly outside the DoE, and it took 3 years for two companion booklets on risk assessment and case studies of environmental appraisal to appear. These reflect the very tentative encroachment of the language and techniques of environmental economics into the policy process.

Although it was too early to evaluate the impact of the initiatives announced in the *Strategy*, probably the most significant body was the Panel on Sustainable Development. The Panel comprises five eminent persons whose role is to provide high level advice by reporting directly to the Prime Minister. Its first report in December 1994 outlined a number of radical proposals, including a shift in taxation away from labour and capital onto resources and pollution, and the phasing out of hydrochlorofluorocarbons (HCFCs). It also announced that it would scrutinise the annual progress reports on the 1990 White Paper, looking at areas where progress has been slow, or targets missed, or where targets appear inadequate. A second new body, the Round Table on Sustainable Development, held its inaugural meeting in January 1995 and its thirty members, drawn from business, environmental organisations, local government, churches and other organisations, planned to meet four or five times a year. Although leading environmental groups were initially suspicious, eventually only Greenpeace boycotted the Round Table (claiming it would be merely a diversionary talking shop). To win the support of other environmental organisations it seems that the government had to extend the objectives of the Round Table from simply helping to clarify the sustainable development agenda and developing a consensus among the main interests involved, to grant it a more active role in policy advice and evaluation (*ENDS Report* 240: 6). Five Round Table groups were formed, covering issues such as energy policy, sustainable transport, and environmental auditing. However, the Round Table's relationship with government got off to a bad start when its first recommendation – the inclusion of an environmental duty on the gas industry regulator Ofgas in the Gas Bill – was rejected by the government (*ENDS Report* 245: 3). Probably least optimism surrounds the final mechanism, the renamed 'Going for Green' initiative, which was eventually launched in February 1995 with the aim of carrying the sustainable development message to local communities and individuals.

The absence of further internal machinery-of-government reforms in the *Strategy* suggests a degree of government complacency. Even if the most optimistic government interpretation of the White Paper reforms is accepted, it could not be claimed seriously that major improvements in co-ordination had been achieved. It was always unlikely that the three new external institutions could play a major role in improving the co-ordination of sustainable development, given their limited terms of reference, lack of formal powers and resources, and their location outside Whitehall.

A cross-sectoral environment agency

Significantly, the sustainable development debate played very little part in the only significant Conservative reform of the machinery of government directly relating to the environment to take place after the publication of *This Common Inheritance*. The White Paper did not herald a rush of environmental legislation: the Environmental Protection Act 1990 pre-dated the White Paper, after which the Environment Act 1995 was the only statute devoted primarily to environmental issues. From 1991 onwards, however, the government was committed to establishing a new Environment Agency, bringing together the functions of the National Rivers Authority (NRA), Her Majesty's Inspectorate of Pollution (HMIP) and the waste regulation functions of local authorities. The planned agency was the only domestic environmental policy pledge in the Conservative manifesto for the 1992 general election; yet its formation was delayed until 1996, indicating the low priority attributed to the environment. The context in which this proposal emerged further illustrates the limited impact of the sustainable development debate on environmental policy-making (Carter and Lowe 1994).

John Major announced the government's intention to create a combined environmental agency in July 1991. Yet 2 years earlier the Secretary of State for the Environment, Nicholas Ridley, had firmly rejected a recommendation by the House of Commons Select Committee on the Environment to set up an inclusive agency, arguing dismissively that 'there is no guarantee that synergy will be gained automatically by amalgamating disparate organisations' (*ENDS Report* 198). Recent administrative reorganisations, including the creation of HMIP and the NRA, encouraged the government to reject further changes in favour of allowing these organisations to establish themselves. A year later, the White Paper adopted a more ambivalent position towards reorganisation, but, with Major's 1991 declaration of support for an agency, in the words of the Environment Committee (1992: xiii): 'In just 2 years, the Government's policy had shifted from outright rejection of the notion of such an agency, to enthusiastic acceptance'.

There is little doubt that the precise timing of the government's conversion to the idea of an agency was motivated by immediate political considerations. It was John Major's first speech on the environment as Prime Minister and, as *The Times* put it, the government was 'eager to be seen wearing fashionably Green clothes' (5 August 1991) in the run up to the general election. Although political opportunism undoubtedly influenced the timing of the Prime Minister's announcement, it was also the culmination of a process of rethinking government policy that had begun several years earlier. By 1991 there was near consensus encompassing government circles, opposition parties and the network of producer and environmental pressure groups in favour of an agency with cross-media regulatory responsibilities for air, water and land. The common denominator within this consensus was one of administrative efficiency:

> It is easier for industry and others to have only one inspectorate to approach; that with separate organizations there is always the risk of some overlap or duplication of effort; and that combining the bodies under one management might lead to greater consistency of approach across all pollution types and environmental media.
>
> (DoE 1990: 18.13)

Having shifted away from traditional end-of-pipe regulation by accepting the principle of integrated (or cross-media) pollution control in the Environmental Protection Act 1990, the case for consolidating the bulk of regulatory functions within one agency became increasingly persuasive.

What was glaringly absent from the debate was any discussion of the role of an agency in enhancing sustainable development. Nevertheless, the legislation setting up the agency does define its objective as contributing to sustainable development. However, reflecting the Conservative government's deregulation initiative, it is also required to take account of the costs and benefits of its proposed actions as well as the costs to industrial sectors and individual companies, before it proceeds.

Discussion

The evidence presented above suggests that, whilst there were a number of changes in the way that the Conservatives approached environmental policy, there remained a marked reluctance to translate the rhetoric of sustainable development into reality. There are a number of explanations for British tardiness.

First, sustainable development is, as Bruce Doern (1993) argues, a 'latent paradigm': although there is now a general agreement over what the concept means in broad terms, sustainable development lacks, as yet, the coherence of earlier (e.g. Keynesian) paradigms and provokes considerable conflict over what it means in practice. The Conservative government undoubtedly adopted a weak definition of sustainable development. In *This Common Inheritance* it was given a limited conservationist flavour – 'the ethical imperative of stewardship' requiring 'a moral duty to look after our planet and to hand it on in good order to future generations' (DoE 1990: 1.14) – which fitted comfortably into traditional Tory paternalistic values. The government view had hardened somewhat by the time of the *Strategy*, which quotes the familiar Brundtland definition of 'development that meets the needs of the present without compromising the ability of future generations to meet their own needs' and throughout displays a clearer understanding of the concept. Nevertheless, rather than adopt the reasonably strong view that sustainable development requires the establishment and maintenance of a minimum environmental stock or capacity (to be defined), the government clearly prefered to accept the inevitability of trade-offs between social and economic preferences and environmental resources (House of Lords, 1995: 10). Moreover, nowhere did the government accept the Brundtland emphasis on equity, either in contemporary or future society, as playing any part in its understanding of sustainable development. It is only at the local level, for example, at the Royal Town Planning Institute and the Local Agenda 21 Steering Group (of local authorities), that ideas of equity and justice appeared prominently on the agenda (House of Lords, 1995: 9). This difference may be in part explained by political differences: while equity is anathema to the Conservative Government, it is far more attractive in local politics where the Labour and the Liberal Democrat parties dominated, particularly among those authorities and organisations enthused by sustainable development. However, British attitudes to sustainable development cannot be wholly explained by political differences, for, in practice, the Labour Government elected in 1997 has also tended to give priority to considerations of economic growth over environmental protection.

Across the political spectrum the central tenet of sustainable development – that environmental protection is a precondition of future (sustainable) growth rather than its antithesis – has failed to dislodge the simple dichotomous belief of economy versus environment. Such a belief underpins the notion that the key role of government is to achieve a balance between the conflicting objectives of production and environmental protection. In this sense the rationale for environmental policy has not advanced beyond Harold Wilson's classic formulation to the effect that '[t]he Polluters are powerful and organised [...] the Protecters, the anti-pollution lobbies, are less organised, less powerful. Therefore the community must step in to redress the balance' (*New Society* 5 February 1970: 209).

Significantly, there has been little discussion in UK government circles of 'ecological modernisation'. As is shown elsewhere in this book, ecological modernisation involves the modernisation (and partnership) of government, business and citizens with the intention of both coping with the challenges of, and seizing the opportunities provided by, environmental issues. This debate is conspicuous by its absence in the UK. It appears, sporadically, only at the level of local government or in specific campaigns organised by environmental groups, such as Greenpeace, aimed at influencing businesses.

Following Weale (1993) it is useful to consider the role of belief systems, as well as institutional and political factors, if we are to understand how new concepts are turned into policy practice. He draws a contrast with Germany where, unlike in the UK, the articulation of principles is integral to the policy style. The UK policy style studiously avoids discussion of principles, so it is hardly surprising that a concept which carries an explicit set of principles as baggage and which has radical implications for the operation of government has been only slowly and grudgingly accepted by the UK establishment.

Belief systems cannot, of course, be isolated from material interests. Capitalist firms will only be converted to ecological modernisation if they perceive it be in their long-term interests, i.e. whether it will ensure their survival and enhance profitability. For example, recycling of waste makes sense only after waste has first been reduced to a minimum; not surprisingly the German packaging industry is just as resistant to encouraging waste reduction as is the British. Nevertheless, it is still legitimate to argue that in relative terms the UK policy elite is further behind the Germans in accepting sustainable development as a dominant belief system.

Belief systems do not change overnight, but the particular institutional and political context seems to have ensured that the transformation has occurred rather more slowly in the UK than in other north European countries which have been subjected to similar international pressures. Thus, in pollution control the British policy elite has found it difficult to grasp the precautionary principle, because it conflicts with the dominant role that natural scientists have played in the process of policy advice (Weale 1993). Unless and until a cause–effect relationship is *proven*, scientific advice is, usually, to do nothing. The tradition of Whitehall departmentalism has led ministers, often encouraged by producer-dominated policy communities, to resist, successfully, almost every attempt to integrate policy-making across departmental boundaries. This institutional barrier effectively prevented Chris Patten from including several more radical proposals in the White Paper. Significantly, the attitude towards sustainable development of the Treasury, which as keeper of the purse strings and with a stake in every department, is the most powerful Whitehall department, was at best apathetic, at worst obstructive (House of Lords, 1995: Chapter 4).

The decline of the environment as a salient political issue during the 1990s provided the broad political context in which the policy elite felt reasonably secure in its non-responsiveness. There was also, of course, the strong ideological distaste within the Conservative Government for intervention, planning and regulation; all of which are integral to sustainable development. The belief in non-intervention meant that, while other European policy elites became active adherents of sustainable development, the government 'lacked the capacity to see the growth potential of the emerging pollution control industry' (Weale 1993: 211) and did not have the desire or ability to plan a developmental intervention to nurture a new technology. Its deregulatory convictions created a climate in which rather than extend protective regulations, existing regulations had to be defended.

This points to a further convergence of institutional and political factors. Economists play an important applied policy role in British government. Consequently, it was not surprising that the government asked David Pearce – an environmental economist – to suggest ways of responding to the Brundtland Report. Clearly he had some impact. The final principle outlined in *This Common Inheritance* declared that the government would seek out ways of using market instruments to encourage producers and consumers to change behaviour, but it was acknowledged that regulation would remain the foundation of pollution control. However, most of Pearce's proposals ended up being tucked away in the appendix – showing Pearce had lost the battle. This can be explained partly by the political context in which the White Paper was written. In 1990, with the environment high on the agenda, the government wished to avoid accusations that by handing responsibility for preventing environmental degradation over to the market, it was effectively dragging its heels. But Pearce had made a powerful intellectual case for market instruments which did appeal to the government in general and Chris Patten in particular. Rather than contributing to a 'rational' debate about the appropriate measures required to deal with environmental problems, Pearce's work became enmeshed in a partisan debate, stirred up by the new right, about deregulation. Little progress was made in implementing these ambitious proposals for market mechanisms. But, having hedged its bets on the market/regulation dichotomy in *This Common Inheritance*, the tone of the DoE 1993 consultation paper on sustainable development was more overtly favourable to economic mechanisms, albeit to reduce costs and controls, not to improve environmental protection:

> Regulation will continue to be necessary in some cases, and is sometimes required by international agreements or legislation. The government is, however, currently reviewing all existing regulatory mechanisms with a view to removing any unnecessary or excessively restrictive controls. For those that remain, the Government will be concerned to ensure that the environmental benefits justify the likely costs (taking account of the risks involved) and that the controls themselves are as streamlined as possible.
>
> (DoE 1993a: 4.19)

The environment had dropped down the political agenda and there was less pressure to introduce new policies. The Major Government had also rediscovered the new right enthusiasm for deregulation. Following the success of Chancellor Lamont's policy of introducing lower taxes on unleaded petrol – with a consequent huge shift in consumption patterns away from leaded petrol – the case for the market could be made with more ideological vigour and greater political acceptability. This was confirmed in the *Strategy* document with the declaration:

> In environmental policy, the commitment is to make use of economic instruments where possible, rather than regulation. Economic instruments seek to make environmental costs explicit and to ensure that people take account of them in making decisions. They can give a continuing incentive for innovation and the development of environmentally friendly techniques.
>
> (HM Government 1994: 3.21)

Yet, despite the publication of a special report on economic mechanisms in November 1993 (DoE 1993b), few specific economic mechanisms were installed. One political obstacle to progress is that economic mechanisms still represent a government intervention in the market and, in practice, ideas like road pricing, carbon taxes and emission charges will generate massive opposition from those individuals and businesses whose behaviour would be restricted. Ministers insisted that any shift in regulatory strategy towards economic mechanisms should be environmentally neutral, should yield significant savings on administrative costs, and should not be seen as affecting business growth adversely (the latter was an overriding consideration given the difficulty in getting British industry out of recession).

Ministerial reluctance was matched by bureaucratic apprehension about the feasibility and effectiveness of economic mechanisms. As a result, general indecision over economic mechanisms combined with political objections to the introduction of new regulatory measures constrained those civil servants working on the strategy document from moving beyond analysis of environmental problems to designing specific solutions for them (Young 1994). Significantly, Treasury inertia impeded progress. The House of Lords Select Committee castigated the Treasury for its lack of enthusiasm for developing economic mechanisms and its complacency in accepting that progress would continue to be a slow 'incremental process' (House of Lords, 1995: 31). Bureaucratic apprehension was, perhaps, justified by the fierce debate that followed the announcement of a new landfill tax that was to be introduced in October 1996. The government encountered almost unanimous objections, on environmental, administrative and revenue-raising grounds, to its proposal for an *ad valorem* tax and eventually agreed to adopt a banded system of taxation. Similarly, although the idea of road taxing has been around for many years, growing enthusiasm for it led only to a limited programme of technical testing of alternative schemes on one stretch of motorway.

The Conservative government faced an acute dilemma: it was ideologically committed to economic mechanisms, but it was understandably concerned about the political consequences of such instruments. The result was a policy vacuum at the heart of government with regard to the design of sustainable policy measures. Thus, a long British tradition of exporting ideas after failing to utilise them in the UK has been maintained. Pearce's ideas have been picked up in many other countries, particularly within the EU; so, ironically, they may be reimported aboard EU policy initiatives. However, the government strenuously resisted the Commission's proposal for a carbon tax on fuels to promote energy efficiency and curb CO_2 emissions. Its opposition can be traced to its unpopularity when it attempted to impose value added tax on domestic gas and electricity bills, which led to a parliamentary defeat and a partial climbdown. UK opposition eventually killed off the proposal for a carbon tax, thus dealing a blow to EU advocates of fiscal measures as a means to reflect environmental costs throughout the economy.

At the same time, seemingly stalled in any market based strategy, the government was very reluctant to strengthen statutory bodies as agencies of sustainable development. New environmental and conservation agencies lacked significant new powers. Local authorities, while seemingly put in the frontline of environmental protection, were discouraged from taking a leading role in pursuing the EU's Fifth Environmental Action Programme. The government's visceral distaste for regulation led it to block the efforts of the European Commission to introduce a directive on strategic environmental assessment, which would have allowed for a formal environmental appraisal of new programmes and policies.

The growing instability in the arena of environmental policy produced some shifting of established administrative traditions under the Conservatives. For example, policy communities have been shaken and, on occasion, penetrated by more pro-environmental groups. International agreements and EU directives have generated legislative responses and a slow trend towards a more standard setting approach. Little of this was the direct result of the principles or practice of sustainable development. The publication of *This Common Inheritance* – the first comprehensive statement of environment policy – hinted at moves towards a more strategic approach, which came to a fruition of sorts in *Sustainable Development*. But in practice the Conservative government generally paid only lip-service to sustainable development: few new policy initiatives emerged; the rhetoric of sustainability remained an almost alien language in the corridors of Whitehall; machinery of government reforms did little to transform the policy process; and the creation of an Environment Agency owed little to the idea of sustainability. However, a few significant shifts have occurred.

In particular, transport policy is one area where the commitment to sustainability produced change. The *Strategy* for the first time acknowledged that unlimited traffic growth is incompatible with the Government's environmental objectives; a conclusion supported by the report of the Royal Commission on the Environment in October 1994, which recommended that the roads programme should be halved, petrol taxes doubled and public transport and cycling encouraged. It is clear that the process of producing the *Strategy*, by generating a confrontation between the transport and environment departments, contributed significantly to this shift and the resulting policy changes. Subsequently, parts of the massive road building programme were put on hold; public spending on roads was cut; road pricing schemes were investigated; and planning policy advice was altered to discourage the building of out-of-town shopping centres. There is still no integrated transport policy and many other policies (such as cuts in railway spending) work against a sustainable transport policy; but these shifts marked a reversal of the 1989 *Roads for Prosperity* White Paper, which viewed traffic growth as a sign of prosperity that should be accommodated by the government.

There is an old lesson here concerning the unintended consequences of policy change. Although the Conservatives were not enthusiastic converts to the sustainable development bandwagon, it gave it official encouragement. Consequently, it provided ammunition to environmental organisations, local authorities and, significantly, the DoE to employ against every pronouncement, new regulation or tax and legislative proposal that failed to give due regard to environmental effects. As the discourse of sustainability gains access to new agendas and audiences, so the cultural, institutional and political constraints on it gaining the acceptance of the British policy elite may be overcome, even though the subsequent slow progress of Labour's reforms suggests it will be a long process.

References

Blowers, A. (1993) 'Environmental policy: the quest for sustainable development', *Urban Studies* 30, 4–5: 775–96.

Carter, N. and Lowe, P. (1994) 'Environmental politics and administrative reform', *Political Quarterly* 65, 3: 263–74.

DoE (Department of the Environment) (1989) *Sustaining Our Common Future*, London: HMSO.

—— (1990) *This Common Inheritance*, Cm. 1200, London: HMSO.

—— (1992) *General Policy and Principles*, Planning Policy Guidance Note 1, London: HMSO.

—— (1993a) *UK Strategy for Sustainable Development*, consultation paper, London: HMSO.

—— (1993b) *Making Markets Work for the Environment*, London: HMSO.

Doern, G. B. (1993) 'From sectoral to macro green governance: the Canadian Department of the Environment as an aspiring central agency', *Governance* 6, 2: 172–93.

The Earth Summit (1993) Graham and Trotman, and Marinus Nijhoff: London, Dordrecht, Boston, MA – official UNCED texts.

English Nature (1993) *Position Statement on Sustainable Development*, Peterborough: English Nature.

English Tourist Board (1992) *The Green Light: a Guide to Sustainable Tourism*, London: ETB.

Environment Committee (1992) *The Government's Proposals for an Environment Agency*, 1st Report, HC 55, Session 1991/2, London: HMSO.

Green Alliance (1991) *Greening Government: the Failure of the Departmental Annual Reports to Reflect Integrated Policy Making*, London: Green Alliance.

—— (1992) *Greening Government 2*, London: Green Alliance.

HM Government (1994) *Sustainable Development: the UK Strategy*, Cm. 2426, London: HMSO.

Hill, J. and Jordan, A. (1994) 'The greening of government: lessons from the White Paper process', *ECOS* 14, 3/4: 3–9.

House of Lords (1995) *Report from the Select Committee on Sustainable Development*, Vol. 1, HL 72, Session 1994–95, London: HMSO.

Local Government Management Board (1993) *Towards Sustainability, A Guide for Local Authorities*, Luton: Local Government Management Board.

Lowe, P. and Flynn, A. (1989) 'Environmental politics and policy in the 1980s', in J. Mohan (ed.), *The Political Geography of Contemporary Britain*, London: Macmillan.

Pearce, D., Markandya, A. and Barbier, E. (1989) *Blueprint for a Green Economy*, London: Earthscan.

Rose, C. (1991) *The Dirty Man of Europe*, London: Simon & Schuster.

Stewart, J. and Hams, T. (1993) *Local Government for Sustainable Development*, Luton: Local Government Management Board.

Ward, S. (1993) 'Thinking global, acting local? British local authorities and their environmental plans', *Environmental Politics* 2, 3: 453–78.

Ward, S. and Lowe, P. (1994) *Adaptation, Participation and Reaction: British Local Government–EU Environmental Relations*, Newcastle upon Tyne: Centre for Rural Economy, University of Newcastle.

Weale, A. (1993) 'Ecological modernisation and the integration of European environmental policy', in D. Liefferink, P. Lowe and T. Mol (eds), *European Integration and Environmental Policy*, London: Belhaven.

Whitby, M. and Ward, N. (1994) *The UK Strategy for Sustainable Agriculture: a Critical Analysis*, Newcastle upon Tyne: Centre for Rural Economy, University of Newcastle.

Young, S. (1994) 'An Agenda 21 strategy for the UK?', *Environmental Politics* 3, 2: 325–34.

8 Paradise postponed

An assessment of 10 years of governmental participation by the German Green Party: 1985–95

Charles Lees

Introduction

Of all the European ecology parties, the German Greens are the most successful and well established. Raschke estimates their 'core' potential electorate to be around 5 per cent of total voters, with an additional 'fringe' potential electorate of around 8 per cent of voters (of which about one third consistently vote Green) (cited in Bürklin and Roth, 1993: 163–5). Such high levels of potential support mean that the party should continue scaling the Federal Republic's 5 per cent electoral barrier to electoral representation.

The German Greens are by no means the oldest ecology party in Europe, nor have they recorded the highest share of votes cast. Both these achievements go to the British Greens. Yet, higher levels of post-materialist value orientation (Inglehart 1990: 163; Padgett 1993; Smith *et al.* 1992; Paterson and Southern 1991; Dalton *et al.* 1984) and a more advantageous political system – including the Federal Republic's arrangements and its system of proportional representation – have provided a relatively benign environment for the Greens in Germany, compared to that enjoyed by their British colleagues. In other words, the 'political opportunity structure' of the Federal Republic – identified by Kitschelt (1986: 57–68) as the combination of coercive, normative, remunerative and informational resources at the party's disposal, as well as the nature of the 'regime's' institutional rules (such as state financing of parties, electoral laws and those reinforcing interaction between governments and interest groups) – has proved to be more conducive to Green mobilisation.

In policy terms, the above conclusion is of central importance, given that this chapter starts from the a priori assumption that, in all areas of public policy, it is not the nature of a given problem, but rather the institutional context within which the problem is addressed that determines the nature and extent of a given solution. As Katzenstein (1987: 7) points out in the German context, 'the interaction between policy and politics is affected primarily by the institutional organisation of power rather than by the imperatives for action inherent in particular policy problems' and environmental policy has not been an exception. As Weale (1992: 79) observes, 'there were elements in the ideological and institutional traditions of German public policy that made certain elements of

ecological modernisation both a legitimising device of public policy developments and a potential source of policy principles'. The nature of these elements
will be discussed in due course, but as Weale (1992: 79) further points out: 'the
ideological and institutional features of German public policy that found ecological modernisation congenial also predisposed the discourse towards specific
elements of the new programme'. This last point is crucial, for this chapter will
argue that the Greens' record in the policy sphere – although quite respectable
when taken in the round – has been subject to institutional restraints that have
encouraged the selective emphasis of certain elements of the green political
discourse. I will argue that, in coming to terms with these constraints, the
Greens have not only shaped the political discourse but, in turn, have been
shaped by political *praxis*.

This chapter, firstly, surveys the extent of Green involvement in subnational
coalitions from the first 'red–green' coalitions (with the Social Democratic Party,
SPD) in Hessen (1985–7) and West Berlin (1989–90) through to the mid-1990s.
Secondly, it assesses the impact that green ideas have had on policy-making; and
considers to what extent the Green Party itself has been responsible for defining
this impact, or whether the peculiarities of the German policy process have
effectively set their own agenda. Thirdly, I outline the extent to which policies
directly associated with the Greens were actually implemented. Given the difficulty, within the confines of a single chapter, of defining and operationalising a set
of common criteria to assess the success of policy implementation across all cases,
the primary empirical component of this section will concentrate upon the
Hessen coalitions and the less-well-known West Berlin coalition that existed
from March 1989 until shortly before the first all-Berlin elections in December
1990. Finally, I analyse the impact of involvement in government on the Green
Party itself, both in the specific cases of West Berlin and Hessen, and in terms of
the general impact upon the *realo–fundi* debate and the organisational principles
of rotation and *Basisdemokratie*, before attempting to reach some degree of conclusion on the overall impact of Green governmental participation upon both the
party and the wider German polity.

The extent of Green involvement in subnational coalitions

The origins of the German Greens are well documented – see, for instance,
Scharf (1994); Markovits and Gorski (1993); Kleinert (1992); Hulsberg (1988);
Bickerich (1985); Muller (1984); Papadakis (1983); Mettke (1982) – and do not
require more than a brief résumé.

The proto-Greens emerged out of the 'citizens initiative' groups of the mid- to
late 1970s. The early years were characterised by internecine struggle between
the ecology movement's 'new left' and conservative wings, until the two wings in
Bavaria decided to unite in the state's elections of October 1978 as *Die Grünen*
(The Greens). This arrangement became known as the 'Bavarian Cooperation
Model' and was to be the template for future co-operation in other *Länder* (Pater-

son 1989: 271–3). Once inside the same organisation, the conservative elements became progressively marginalised and the proto-party began to assume its familiar left-libertarian/post-materialist character (Markovits and Gorski 1993: 192–7). The proto-Greens also campaigned in the 1979 elections of the European Parliament,[1] winning a respectable 3.2 per cent of the vote (Mintzel and Oberreuter 1992: 542). However, although many local green parties formed during this period, others did not contest local elections until after the formation of the national party in 1980. For instance, with the exception of *Alternative Trier*, no local elections were contested in the Rhineland–Palatinate until 1984 (Scharf 1994: 64–6). Thus, the formative experiences of local green parties did not always follow a set pattern. The big breakthrough for the Greens came in March 1983, when the national party entered the *Bundestag* for the first time, having won 5.6 per cent of the vote in the national elections (Padgett 1993: 28). Not only did this raise the profile of the Greens per se, but it forced the SPD to finally address the post-materialist agenda (Grant *et al.* 1988: 246–7). With the Greens continuing to record respectable levels of support – including 8.2 per cent in the 1984 European elections (Kolinsky 1990) – it was only a matter of time before a state election would produce both a distribution of party weights that would make a formal red–green coalition mathematically possible, and a configuration of local political personalities/conditions that would make it feasible.

The first coalition was formed in Hessen in 1985 as the culmination of 2 or 3 years of tentative co-operation between the Hessen Greens and the local SPD, following the 1982 elections. The local Greens – led by a group based around Joschka Fischer, ex-SPD politician Karl Kerschgens and expatriate French student activist Daniel Cohn-Bendit – were broadly on the *realo* wing of the movement, which made them relatively well disposed to co-operation with the SPD.

The 1989–90 red–green coalition in West Berlin was more tenuous. The local green party – the *Alternative Liste* (AL) – was widely considered one of the more fundamentalist in the Federal Republic. Moreover, the local SPD – although possessing a strong new-left element – remained tainted in the eyes of many green activists from the late 1970s and early 1980s, when the ruling SPD–FDP (Free Democratic Party) city government had been enmeshed in property scandals. This period also coincided with the rise of the squatter movement and the polarisation of political discourse within the city. Thus, within West Berlin's green milieu, there was a great deal of suspicion that had to be overcome. Nevertheless, the coalition took office amid high expectations. Unfortunately, the coalition collapsed amid mutual recriminations (including the AL tabling a vote of no confidence in SPD governing mayor Walter Momper) shortly before the first all-Berlin elections of December 1990.

Despite the mixed experience until then of Green–SPD co-operation, there was an implicit understanding on both sides that co-operation between them was a learning process, and that relations between the Greens and the established political parties were becoming more institutionalised.

The institutionalisation of the Greens has been accelerated by six factors. First, following unification, the merger of the more moderate *Bundnis 90* and 'eastern' Greens (Bürklin and Roth, 1993: 165–72) with the Greens in the 'old' Federal Republic, has resulted in an overall moderation of both the Green voters' and Green membership's ideological profile. Second, the generational cohort from which the movement in the West originally arose has become older and, perhaps crucially, more established and integrated. Third, the growth of bureaucratic structures and hierarchies within the party has meant that an increasingly pragmatic and, I would argue, office-seeking elite has been able to move the party away from the previously rigid *fundi* position of the 1980s. A fourth development has been to what extent the agenda around which the Greens have mobilised, has been co-opted by the other parties. Whilst this has lessened the distinctiveness of the party, it has made them a more viable coalition partner. Fifth, with the decline of the liberal FDP, the Greens have become more decisive within the coalition equation. Finally, much of the stigma that had previously attached itself to the Greens prior to unification, has now been transferred to the reform-communist Party for Democratic Socialism (PDS).

Because there is less political risk in other parties co-operating with them, Green involvement in subnational coalitions has become an increasingly common characteristic of the party system in the Federal Republic. Not only have the Greens participated in a coalition with the SPD in Lower Saxony between 1990 and 1994, but also in the eastern German state of Saxony–Anhalt (a minority administration tolerated by the PDS) between 1994 and 1998. They remained in office until 1999. Moreover, the red–green coalition running Hessen from 1991 was actually re-elected into office in the state elections in early 1995. Finally, the increasing acceptability of the Greens to the bourgeois parties has been demonstrated by the two so-called 'traffic light' coalitions (with the SPD and FDP) in Brandenburg (1990–4) and Bremen (since 1991), as well as the formation of the first 'black–green' coalition (with the Christian-Democratic Union, CDU) in the Westphalian town of Mülheim. Many commentators consider it only a matter of time before such a coalition is formed at *Land* level (*Der Spiegel*, 1 August 1994: 24); it demonstrates the extent to which the Greens can now be regarded as a mainstream party at the subnational level.

An assessment of Green ideas on the policy-making process

The SPD's 'Blue Skies over the Ruhr' programme of 1961 was an early manifestation of environmentalism in post-war Germany (Paterson 1989: 267–8). However, opposition from intermediary organisations and institutional breaks within the German policy-making process would continue to impose constraints upon the environmentalist agenda.

Traditionally, the German state has been held to transcend the vagaries of partisan rivalry and represent what Dyson has described as 'a wider, universal and ethical community' (Dyson, 1982: 78). German administrative culture has been steeped in the traditions of Roman law, with an emphasis upon the cipher of the impartial and (crucially) expert administrator. It was this lack of legitimacy surrounding party government in Germany that prevented the emergence of a robust democratic culture and finally undermined the Weimar Republic and led to the 'totalitarian partnership' between state, society and party that characterised the Third Reich (Broszat 1981: 348). The post-1945 settlement integrated political parties into the centre of the governmental/administrative nexus of the new Federal Republic. However, this was not a clean break with the past, given that the administrative culture within the permanent civil service was and remains deeply rationalistic and expert oriented. Thus, the Federal Republic is characterised by a duality between the conflicting ethos of the contemporary 'party state' (*Parteienstaat*) and that of the residual 'administrative state' (*Beamtenstaat*).

The importance of the political parties within the administrative process meant that, by 1972, the established parties had so successfully penetrated the civil service that over half of the senior posts (state secretaries, heads and departmental heads of division) at the state and federal level of the civil service were staffed by them (Paterson 1982: 105). This made it very hard for any new party – especially an 'antiparty party' – to break into the policy process. This went beyond a question of restaffing a particular department and manifested itself in discursive terms: administrative values have permeated internal debate within the established parties and reinforced the existing consensus between them on the substantive issues of the state. This was most evident in states in which one party had been dominant (for instance in several SPD governed *Länder*) and local parties become permeated by the administrative/legal discourse. It was in such states – especially during Schmidt's chancellorship – that the proto-Greens provided a political home for those both inside and outside the SPD with a left-libertarian value orientation. In this sense, Willy Brandt was correct in calling the Greens the 'SPD's lost children' (Markovits and Gorski 1993: 81).

However, the Greens did have an impact upon policy-making during the 1980s and early 1990s. This came about both in discursive terms and as a pragmatic response to the success of the Greens at the ballot box.

In discursive terms, the Federal Republic's policy-making process did not constitute an absolute block on the Green agenda but rather shaped it and limited its left-libertarian impact. Despite the emphasis upon the rational and authoritative cipher of the civil servant or *Beamte*, German policy-making is comparatively well disposed towards its own 'greening'. Weale's comparison of the policy-making communities in Germany and the UK, demonstrates how German policy has entered its 'recovery phase' (Müller in Weale 1992: 68) since the 1970s, to the extent that Weale believes that 'within Europe Germany has earned for itself the title of an environmental leader' whilst Britain remains a 'laggard' (Weale 1992: 69–70). Weale sees the reason for this as lying in the two countries contrasting policy styles, stating:

> The German policy style [...] is consistent with the operation of a rigid constitution. A programmatic statement of general principles is seen as an essential prologue to legislation and policy development, a tendency that is

probably reinforced by the practice of coalition government in which political parties of different ideological persuasions have to come to some agreement on the running of government [...] The policy community is usually wider in Germany than in the UK. The constitutional formalism of German policy making means that the courts play an important role in the setting of standards.

(Weale 1992: 81–3)

Thus, Weale identifies the paradox at the heart of the Greens impact upon the policy-making process. On the one hand, there are centripetal forces within the administrative apparatus – both in terms of the dominance of managerial values (and therefore a cross-partisan consensus) within the mainstream parties and the need for the Greens to find some form of *modus vivendi* with the SPD – that have constrained the ability of the Greens to implement a left-libertarian environmental agenda. Conversely, the fact that the German policy community places such an emphasis upon expert opinion gives it width: with the potential to grant access to new actors if in possession of such expertise. As Geoffrey Roberts observes, in the Federal Republic

where there is a large degree of consensus, it tends not to be a static consensus; it is agreement which adapts, develops, changes, and – most importantly – it is a consensus that the parties shape and modify by their inputs into the policy making process.

(Roberts 1989: 53)

Finally, the Greens – like all citizens of the Federal Republic – retain the option of resorting to the courts: both in order to challenge the actions of corporations or the state as well as to question the constitutional position of statutes against the Federal Republic's Basic Law. These rights were used to good effect in the campaign against nuclear power during the late 1970s and early 1980s when courts were willing to withdraw or delay the granting of construction permits for nuclear plants. As a result, by 1984 the average construction delay for such plants had risen to 42.4 months (Kitschelt 1986: 80).

The other main source of pressure open to the Greens in order to affect the policy-making process lies in their continued electoral success. This pressure operates upon two fronts. First, it serves to mobilise and maintain support for environmental issues in general and, therefore, to sustain the diffuse pressure upon the policy-making process described above. Second, a strong Green representation in a *Land* legislature has occasionally resulted in them holding the balance of power and thus extracting concessions from one of the established parties, in particular the SPD. However, it must be remembered that these have been concessions extracted during the process of coalition bargaining concerning anticipated policy outcomes over the life of one or more legislative periods. In the early years in particular, coalition bargaining has highlighted the constraints imposed by the Greens' outsider status: not only in relation to the structures of the established policy community but also in terms of a certain degree of political *naïveté* as to what they and their coalition partners were really able to deliver.

Nevertheless, by the end of the 1980s all the established parties were at least paying lip-service to green ideas. Indeed, it was the ruling CDU/CSU (Christian-Social Union) FDP coalition in Bonn that oversaw Germany's 'recovery period' described earlier. In discursive terms, the imposition of the *Vorsorgeprinzip* (precautionary principle) into policy discussion has freed up what had been a rigidly scientific-administrative style and required officials to consider potentially far-reaching and expensive measures, even where a degree of uncertainty amongst expert opinion remained as to their necessity and efficacy. This has helped to bring about domestic policy successes – such as the 1983 Large Combustion Plant Ordinance and the *Grüne Punkt* recycling scheme of 1992 – and official pressure on German's neighbours through the European Union and other international organisations (see Weale 1992; also Grant *et al.* 1988). The adoption of these measures at the level of the European Union must be considered one of the green movement's most substantial successes to date.

The irony is that, for the Greens, the established political and administrative structures' partial adoption of their agenda served to undermine their own identity. As Markovits and Gorski (1993: 236) observed,

> virtually all issues, be they women's rights or ecology, peace or nuclear energy, had been appropriated by all West German parties in the course of the 1980s. The most convincing evidence for the Greens' success was furnished by the party's loss of uniqueness.

The Hessen model

The red–green coalition of 1985–7 (and to a lesser extent 1991–5: the coalition was re-elected in February 1995) is considered by many to be the template for such coalitions: both in other states and to some extent at the national level. It is certainly the most documented (Scharf 1994; Markovits and Gorski 1993; Padgett 1993; Kleinert 1992; Hulsberg 1988). One reason for this has been Hessen's strong *realo* tradition, which has meant that local politicians such as Joschka Fischer have also been leading figures nationally. In addition, Hessen's election system – with its 5 per cent election barrier – is an analogue of the national system as a whole: thus allowing extrapolation up to the national level. However, the reasons for the importance of the Hessen case have been threefold: it was the first formed at *Länder* level; it has since been the longest-lasting; and the first to be re-elected.

The red–green experiment (as it was branded by its critics) of 1985–7 is best remembered for Joschka Fischer's tenure as the first Green Minister for the Environment.[2] Moreover, this is a good example of both the facilitating and constraining nature of the German policy-making process. Fischer was firmly on the *realo* wing of the party and this was reflected in his strategy of concentrating on the stricter implementation of existing legislation (*nach Gesetz und Recht*). In this he was quite successful: for instance, with regard to the chemicals industry, he forced companies – particularly Hoechst AG – to implement new instrumentation in order to meet lower permitted limits of industrial discharges into the River Main. Given the sheer economic and political clout wielded by the Hoechst concern in the state of Hessen, the success of Fischer's Ministry in enforcing these changes had a seismic impact upon future expectations.

Despite this high profile role, Fischer encountered staffing problems within his new Ministry and was forced to appeal to the Administrative Court in Wiesbaden in order to be granted permission to establish a hypothecated personnel advisory board, to sidestep the existing (SPD dominated) boards in order to address the problem (Grant *et al.* 1988: 253–5). As will be discussed later, the problem of staffing of Green ministries was not a problem confined to the Hessen case, but was to some extent evident wherever they came into office. In that instance, there had been explicit threats of disinvestment in the state from Dr Joachim Langmann of Merck, then president of the employers' umbrella organisation.

Langmann's threats did not eventually come to much and the period of 1986–7 – with the Greens polling 10.4 per cent in Hamburg and over 8 per cent in the national elections – seemed to augur well for the coalition. However, the tension between the *realo* and *fundi* wings of the Greens, and the left and right wings of the SPD eventually began to spill over more and more into the workings of the coalition. One of the reasons for this was that many of the more contentious issues that the coalition had to deal with – such as what to do with the NUKEM and ALKEM projects in the town of Hanau[3] – were put off rather than tackled head-on. Although this bought time for the coalition, by the spring of 1987 it had broken down in bitter recrimination over practically every aspect of their programme. The descent into 'red–green chaos' was exploited by the SPD and Greens' conservative opponents. Ironically, the 1987 state elections in Hessen did prove to have some relationship with the nature of government at the national level, in that they produced a Bonn-style CDU–FDP coalition (Markovits and Gorski 1993: 191). Nevertheless, the Hessen experience was the first step towards a normalisation of the Greens' relationship with other parties and, by implication, the German state. Moreover, it spelt the end of the growth-oriented consensus in government.

Whilst the red–green coalition of 1985–7 is considered to be the seminal example of this type of coalition arrangement, it can therefore be taken as the starting point for studying subsequent coalitions. The coalition of 1991–5 is important because it is both under-researched and lasted much longer than its predecessor. Moreover, in February 1995, it was the first red–green coalition to be voted back into office. The division of legislative seats and, consequently, ministerial portfolios are set out in Tables 8.1 and 8.2.

With reference to Tables 8.1 and 8.2 it is apparent that – unlike in West Berlin in 1989 – the Greens did better than the SPD in the numerical division of portfolios. Whereas the SPD's share of portfolios represented a ratio of 7.3 legislative seats to each portfolio, the Greens' share was 6.5:1. Much has been made of the

Table 8.1 The municipal elections of 19 February 1995 in Hessen (in percentages)

Party	% of (second) vote	Number of seats
SPD	38.0	44
CDU	39.2	45
FDP	7.4	8
Bündnis90/Die Grünen	11.2	13
Others	4.1	0
Total	100	110

Source: *Süddeutsche Zeitung*, 21 February 1995.
Note: Turnout: 66.6 per cent. Note that the turnout throughout the period 1981–9 fell 5.7 per cent overall. This is in keeping with the trend towards increased non-voting noted by Paterson and Southern (1991) amongst others.

importance of the Greens' increased share of the vote in maintaining the coalition – see, for instance Green (1995: 153); the Greens' portfolio share would appear to represent the payoff for this electoral success. However, within a matter of months, the Hessen Greens had suffered a massive collapse in morale. Even though the coalition remained popular with the voters, the *Fraktion* in Wiesbaden was perceived as being out of touch with the *Basis*, bereft of ideas and accident-prone.

The problems of the Hessen Greens can, arguably, be directly or indirectly blamed on the decision of local political godfather Joshka Fischer to move to Bonn and become *Fraktionsvorsitzender*. The move to Bonn meant that the local party was denied his undoubted political skills and there was no obvious succes-sor. Moreover, the two Green ministers in the coalition were perceived to have frittered away their political advantage. This led *Der Spiegel*, for instance, to talk about the Greens' difficulties as a 'mid-life crisis'.

The Hessen Greens' troubles were essentially two-fold. First, they were

Table 8.2 Composition of SPD–Bündnis 90/Die Grünen cabinet (Hessen 1995)

Ministry	Minister	Party
Interior, Agriculture, Forestry and Nature Protection		SPD
Finance		SPD
Culture		SPD
Science and Arts		SPD
Economics, Traffic and State Development		SPD
Work, Women and Social Order		SPD
Justice and European Affairs	Iris Baul	Bündnis90/Grünen
Environment, Energy, Health, Youth and the Family	Rupert von Plottnitz	Bündnis90/Grünen

Source: *Koalitionsvereinbarung für die 14. Wahlperiode des Hessichen Landtags zwischen Bündnis 90/Die Grünen & SPD 1995–99* (Bündnis 90/Die Grünen 1995).

becoming very much like the other political parties in both style and, increasingly, substance. This led to disillusionment amongst young voters. As one young academic put it, 'the automatic impulse to vote Green is broken'. Although the majority of Green voters remained true to the party, the PDS was able to mobilise support at the margins of the Greens core support. This tendency was not helped by Fischer's insistence that the Hessen Greens take over a 'classical' ministerial portfolio like that of Justice. This may have made sense in terms of demonstrating the Greens' ability to govern, but not play well with the party's *Basis*. Despite the fact that the Justice portfolio was taken by Rupert von Plottnitz (whose credentials as a former lawyer for the Red Army Fraktion seemed ideal for the job), the Ministry was perceived as not being an appropriate area for a Green politician to be interested in. This was despite a commitment in the coalition agreement to undertake a programme of reforms, including the reduction of state surveillance of citizens, data protection measures, increased use of non-custodial sentences and similar measures.

Second, the trade-off for getting the Justice Ministry was that the Greens had to accept the creation of a 'superministry' for the Environment, Energy, Health, Youth and the Family. This exposed Minister Iris Baul to conflicting political demands that proved desperately hard to reconcile. On the one hand, the 'post-materialist' agenda of environmental protection placed a premium on a critique of existing biases toward production and consumption. On the other, whilst an explicitly Green policy for Health, Youth and the Family can be envisaged, these policy areas are traditionally the domain of statist and production–consumption-oriented solutions. Emphasis has always been on delivery systems to address specific problems rather than on a holistic approach. Iris Baul was faced with trying to reconcile these demands. In addition to this strategic problem, Frau Baul never managed to establish a working relationship with her State Secretary. She resigned in September 1995.

The Red–Green coalition in West Berlin 1989–90

At the time of the January 1989 municipal elections in West Berlin the obvious need to ecologically modernise the economy was not only recognised by the Berlin AL but also by the mainstream parties. For instance, in their election literature, the Berlin SPD stated that 'protecting the environment is also a growth industry, protecting the environment creates jobs' (*SPD Länderverband Berlin* 1988: 7). Thus, although the SPD was still constrained by the traditional materialist discourse and saw the maintenance of jobs as paramount, there was some commonality with the Greens as to the problems the city faced.

With regard to specific problem areas, both parties identified in their election literature the most pressing areas of environmental concern as those of waste disposal, housing, transport and energy policy. However, the AL's proposals were both more radical and less concerned with the impact of their policy proposals upon economic growth and employment. This contrasted with the more technical-administrative style of the SPD (*Alternative Liste Berlin* 1989: 3; *SPD Ländesverband Berlin* 1988: 9).

In purely numerical terms, the AL did not do as well as the SPD in terms of the division of portfolios. The January elections resulted in the election to city hall of 17 AL representatives and 55 from the SPD (*Statistisches Landesamt Berlin* 1989). However, the AL received only three out of the fourteen cabinet posts on offer: a ratio of legislative seats to cabinet posts of almost six to one (the SPD ratio was just five to one; and in addition they appointed the governing mayor). This reflected the disproportionate weight the AL placed upon the environment portfolio. Moreover, the coupling together of city-development and traffic policy with environmental protection (rather than with construction and housing, for example) was considered a hard fought and essential victory for the AL in the teeth of SPD opposition.

The coalition took office to a mixed reception from West Berlin's intermediary organisations. The employers took a wait-and-see attitude, whilst the reaction of the local trade unions ranged from the almost outright hostility of the construction workers' union to the more supportive stance of the education and science union, which also had quite strong links with the AL (Schools Minister Sybille Volkholz had been a union official). Inevitably, there was a marked correlation between the stance of individual unions' and the priority given to their producer interests by the new coalition. For instance, the coalition was committed to an early expansion of nursery education and crèche facilities, whilst its attitude towards the building sector was more ambiguous.

Civil service reaction was also mixed. Given the 8-year incumbency of the previous administration, a significant proportion of the civil servants were CDU or FDP placemen, especially amongst the top tier of permanent officials. Under German law these officials enjoyed security of tenure and could not be sacked; although they could be granted indefinite leave (*Ruhestand*) at the public expense. Whilst it could be expected that some – especially the more senior and/or well-connected officials – would retire or move on, any incoming minister wanting to make major changes was confronted with the cost of keeping superfluous civil servants on the payroll and finding staff of sufficient credibility and expertise to replace them. Given these constraints there was a danger of 'implementation drag' on the part of recalcitrant officials.

However, this was not the case in the City Development, Environmental Protection and Traffic Ministry. There were three reasons for this. First, as the Ministry was only set up in 1981 it had not developed the rigid policy networks (based on producer interests) found in some of the more established ministries. Second, it was staffed with younger officials who could be assumed to be inherently better disposed to ecological modernisation policies and therefore more open to innovation. Third, the Ministry had previously been close to the FDP and – as they had failed to pass West Berlin's 5 per cent electoral barrier and were not represented in the legislature – any latent opposition that did exist amongst permanent officials lacked a parliamentary focus. As a result, the new Environment Minister and her staff were confident they could successfully implement the agreed programme.

The programme placed great emphasis upon greater transparency in what they called the city's 'planning culture'. Other specific commitments included: new controls over air emissions and the encouragement of new clean-air technologies; removal of lead piping in the city's water provision; imposition of state-of-the-art technology within the water industry; improved river management; imposition of an integrated and coherent waste-disposal policy in co-operation with other *Länder*, a freeze on the new development of 'green field' sites; the development and imposition of a 'sparing, rational and socially sustainable' system of energy provision and use (including a new energy tax, modification of existing laws on energy use, modification of pricing system and decentralisation of energy production); a long-term plan to completely reopen the city's railway system (*S–Bahn*), the expansion of public transport through extension of bus lanes, reduced waiting times, reduced prices including introduction of a cheap all-inclusive travel card ('*Umweltkarte*').

Three main observations can be made about the coalition's proposed programme. First, with the exception of the reduction of prices for public transport, most of the programme required implementation over the medium to long term. This meant that political costs – such as funding that would have otherwise gone on construction and service provision for instance – were felt immediately, whilst the benefits were of little short-term use: given that they were deferred and/or intangible or counterfactual in nature (in other words the greater costs of non-implementation would not be apparent as they would presumably have been avoided). The perceived lack of results was a source of frustration for both coalition partners, as expectations from both the SPD's supporters and the AL's *Basis* proved a constant source of pressure upon the maintenance of the coalition.

Second, the policy document was in many ways the product of two distinct and often contradictory discourses. The left-libertarian/post-materialist link between environmental policy and wider societal power structures was plainly evident; for instance, in the greater role allocated to local communities in future planning decisions and the opening out of the policy community itself (through more horizontally-structured 'working groups' and the inclusion of non-governmental organisations and self-help groups). At the same time, however, a more statist approach remained evident and, arguably, could hardly have been avoided: given that the coalition remained reliant upon the German 'administrative state' (*Beamtestaat*) to, for instance, recodify law, administer increased subsidies to public transport and collect ecotaxes. It was this tension between the two approaches that is at the heart of the 'old left'/'new left' dialectic (Inglehart 1990; Van Deth and Geurts 1989; Eckersley 1989).

Finally, there was a heavy reliance upon bureaucratic instruments – such as judicial review, state regulation and subsidy – with economic instruments (such as ecotaxes and pricing) taking a secondary role (see Table 8.3). This was especially the case in those sectors that constitute indivisible public goods (such as air and water). The reliance upon bureaucratic instruments represented continuity rather

Table 8.3 Six environmental policy sectors and the choice of regulatory instruments (bureaucratic and/or economic), West Berlin, 1989

Sector	Instruments
Air	Bureaucratic
Water	Bureaucratic
Waste	Bureaucratic/economic
Land	Bureaucratic/economic
Energy	Bureaucratic/economic
Traffic	Bureaucratic

Source: *Berliner Koalitionsvereinbarung zwischen SPD und AL vom 13. März 1989* (SPD Ländesverband 1989).

than change in the style (if not the content) of policy-making in the Federal Republic. Despite the presence of the AL in government, it was clear that their impact upon the structures of public policy was limited. Moreover, with the exception of peripherals such as the *Umweltkarte*, the coalition did not survive long enough to make any long-term impact upon the content of policy. Ultimately, the coalition's main impact was cognitive. As one SPD official put it, 'the coalition's main achievement was that it existed' and to some extent redefined what was achievable within a social market economy. But it all seemed a far cry from the Greens' early visions of utopia.

Analysis of the impact of governmental participation on the Greens

In 1988, Werner Hülsberg – no stranger to the internecine struggles of the Greens – wrote:

> The Green party lacks an inner equilibrium. The excitement of an unexpected electoral success is followed by inner-party strife and fierce factional battles. But those fiercely-fought battles are without consequence because everyone is aware of the over-riding need for unity. At the same time the weakly-developed party apparatus is incapable of organising any kind of internal repression. The party, in this sense, is still a kind of electoral pact. This situation will not change overnight. Majorities at party conferences are not an expression of any kind of long-term development but rather the expression of the mood of the moment [...] The development of political strategy takes place in an empirical manner, under the Damoclean sword of the five percent hurdle, what Joschka Fischer calls "the pressure of actual circumstances".
>
> (Hulsberg 1988: 212)

It is self-evident that the Greens have undergone a profound metamorphosis since these words were written, but it is less clear as to why this change has come about. In general terms, how are we to define such structural and programmatic changes within parties and why do they take place?

There is a rich vein of party-system literature addressing this question– see, for instance, Michels (1970); Lippman (1914); Panebianco (1988); Katz and Mair (1992). However, for the purposes of this paper, Harmel and Janda's (1994) definition of party change is sufficient. Harmel and Janda regard party change as being the function of three factors: external shock (e.g., a bad election performance), leadership change and change in the dominant coalition. Of the three, data from both Germany (excluding the Greens) and Britain between 1950 and 1990 indicate that external shock, whilst important, is the least decisive in bringing about changes in the organisational and issue positions of individual parties (Janda et al. 1995: 172–3). Thus, where changes have taken place, they have been more likely to be the result of changes at the very top or within the dominant coalition of the party.

This is only to some extent the case with the Greens: not only because they do not have a formal party leader in the style of the more established parties, but also because pressures from the exogenous world remain central to the process of change. Nevertheless, that a process of programmatic and structural renewal has taken place has surprised many observers. Given that the unstable majorities of grass-roots activists at party conventions have tended to produce extreme and often contradictory resolutions about the party's strategy and programme, many observers had concluded that fundamental renewal of the party would only lead to further schism and dissolution on the left. However, that such change did come about was indicative of the growing institutionalisation of the Greens: both in terms of its internal structures and in its relations with the exogenous party system.

The failure of the 'western' Greens to pass the 5 per cent barrier in the 1990 federal elections sped up the process of union between them and the 'eastern' Greens. Given that such external shocks have been rare for the Greens at the national level, this would appear to be fully in line with Harmel and Janda's hypotheses. However, at the subnational level, the Greens suffered external shocks that have not only brought about organisational and programmatic change, but have also been central to changes in the dominant coalition itself. For instance, in West Berlin, hostile media coverage of the crises and eventual collapse of the red–green' coalition led the AL to tone down its policy of transparency in internal debate and adopt a more disciplined approach. At the same time, the practice by which Green ministers had to refer all major decisions back to their *Basis* was phased out. In Hamburg, the uncompromising stance of the *Grüne Alternative Liste* (GAL) was widely believed to have led to the GAL's vote share dropping from 10 per cent to 7 per cent in May 1987 and fuelled the strategic debate at the national level (Hulsberg, 1988: 213–4). In Lower Saxony, the ambiguity of some elements in the local party towards political violence was regarded as a major factor in the failure of the left to unseat the CDU–FDP coalition the following June and led to a review of the local party's hardline approach (Markovits and Gorski 1993: 211–7).

Such electoral disappointments at the subnational level spurred the processes of change within the party but were not decisive in themselves. Indeed, one could argue that such changes were inevitable: given the 'pressure of actual circumstances' that Fischer described. Petra Kelly's original description of the 'antiparty party' for whom 'parliament is not a goal but a strategy' (Markovits and Gorski 1993: 121) remained a rallying call, but even the most mundane day-to-day problems associated with political praxis tended to generate the kind of solutions that served to distance the Greens from this vision and lead to the development of an increasingly recognisable hierarchy within the party.

Harmel and Janda's hypothesis is borne out to the extent that the pressures for change were not just of a diffuse and technical nature but also the direct result of the actions of individuals and coalitions within the party. This was particularly true as the party's internal strife between the *realo* and *fundi* wings became more entrenched. Over time, ad hoc divisions of responsibility (and by implication, chains of command) sprung up within both the parliamentary *Fraktion* and in the wider party in response to the power struggle. As personalities – such as Joschka Fischer on the *realo* side or the more *fundi* Jutta Ditfurth – became publicly associated with a particular political stance, hierarchies inevitably grew around them to deal with the logistics of *de facto* leadership. These hierarchies tended to be divided within the party's executive (where the *fundis* were in the ascendant) and parliamentary *Fraktion* (were the *realos* were stronger), thus exacerbating the division between *Fraktion* and party. In turn, as these debates became increasingly public – and fired by external shocks such as those described earlier – the normal procedures of grass-roots democracy broke down. Policy often appeared to be made in the media arena rather than via internal debate.

Ultimately, the processes of change became unstoppable and the *fundis* lost the ascendancy that they had enjoyed in the early and mid-1980s. There appear to be three main reasons for this. First, as already described, there were technical imperatives associated with the praxis of party politics that spurred on the growing institutionalisation of the Greens. Second, as the *realo–fundi* debate polarised between the *Fraktion* and the party executive, the former's superior resource base (such as access to the media, office facilities and funding) began to exert a decisive influence upon the Greens' external image and internal discourse. Moreover, in the absence of a mass membership, the party as a whole relied upon state funding for its representatives in order to survive, which in turn strengthened the *realo's* position. Nevertheless, it was to be the ordinary party membership who finally ended the 'battle among the mullahs'. Many in the increasingly alienated *Basis* perceived the struggle as being just as much about personalities as politics, a fact that led to the emergence of a third centrist faction called *Aufbruch* (fresh start) in the late 1980s. Consequently, an alliance was struck between *Aufbruch* and the *realos* at the party's federal convention in December 1988 and the incumbent (*fundi* dominated) executive was replaced by a more representative committee (Padgett 1993: 178). The *realo–fundi* conflict was effectively over and with it the explicitly 'antisystem' phase in the party's development. The Greens were moving from the political margins towards becoming a potential party of government.

Conclusion

By the mid-1990s the Greens had decisively made the transformation from an explicitly 'antiparty party' to a player within the political mainstream. Indeed, with the decline of the FDP – which at present appears to be terminal – they have been increasingly accepted as the 'third force' within the German party system. State elections in Bremen and North-Rhine Westphalia on the 14 May 1995 vividly bore out this trend. In both states, the voters not only rejected the FDP but the Greens made gains at the expense of the SPD. In Bremen, the Greens slightly increased their vote to 13.5 per cent, whilst the SPD dropped to 33 per cent. In North-Rhine Westphalia, the Greens doubled their vote to 10 per cent, whilst the SPD won 47 per cent (*Süddeutsche Zeitung*, 15 May 1995). Whilst the SPD retained an absolute majority of seats in North-Rhine Westphalia, the loss of vote share in their heartland was seen as a further signal from the electorate that co-operation with the Greens was no longer an option of last resort for the SPD, but increasingly an essential prerequisite to holding office in many states. Indeed, one headline in the German press stated that the Greens had now taken over the function of 'king maker' previously enjoyed by the FDP (*Handelsblatt*, 16 May 1995).

In reality, the notion that the Greens are now the 'king maker' within the German party system is at least premature and probably inappropriate. This is because the FDP remains programmatically compatible with both parties, having both an economically and socially liberal component within its ideology which could be emphasised depending on who was its senior coalition partner. For instance, during the 1969–82 Social–Liberal coalition, the FDP emphasised the 'social' side of its ideology and, after the formation of the coalition with the Union parties in 1982, its economic liberalism became more dominant. This Janus-like quality traditionally lent credibility to the FDP's ultimate threat to any coalition partner: to form a coalition with the opposition. However, apart from the isolated case of Mülheim (which fuelled much of the present speculation in the media), it is hard to envisage a 'black–green' coalition in the near future and certainly not at the national level. Thus, whilst Joschka Fischer and others may hold up the possibility of co-operation with the CDU/CSU, it is more as a bargaining position in the Greens' manoeuvres with the SPD than a credible threat at present.

Red–green coalitions, on the contrary, have become commonplace at the *länder* level. In addition to those mentioned earlier, there have been red–green coalitions in North-Rhine Westphalia (1995–), and Schleswig Holstein (1996–). Moreover, since 1998, a red–green coalition has been in power at the national level.

That the Greens have both stabilised their vote share and become an established feature of the party system may have surprised some observers but is not without precedent. Indeed, the SPD itself underwent the same process earlier this century (Padgett 1994a,1994b). However, what is perhaps striking is the speed at which this process has taken place, despite the Greens' lack of a clear social cleavage around which to mobilise. This has led some observers to wonder if a 'post-industrial' cleavage now exists or, alternatively, that cleavage structures are no longer relevant to the modern German party system (Bürklin and Roth, 1993: 160–2).

It would be rash, however, to deny the persistence of the old materialist, social-political divisions or speculate as to their continuing salience in the future. As this chapter has demonstrated, the German party system has adapted to accommodate the Greens rather than been transformed by them. Indeed, the only party that has unequivocally been a casualty in the process of partisan realignment and electoral dealignment in the Federal Republic is the FDP, who lack a stable cleavage (or even a distinct milieu like the Greens) around which to mobilise, and have therefore been more exposed to the most adverse effects of such changes. Conversely, whilst the two big *Volksparteien* have seen some slippage in their vote, they remain the major players within the party system and the Greens have had to deal with this fact.

To conclude, the constellation of political, administrative, cultural and economic variables in the Federal Republic have provided the German Greens with an opportunity structure that has proved more conducive to left-libertarian and ecological politics than elsewhere in western Europe. Moreover, the Federal Republic has not only become one of the most environmentally-enlightened societies in the western world but is also pushing hard (particularly through the European Union) for her neighbours and trading partners to do the same. Much of the credit for this must go to the German Greens.

However, the Greens could not have hoped to remain aloof from the pressures of political praxis and this opportunity structure has performed a constraining as well as enabling role upon the party and the wider green agenda. In this sense, any realisation of the more utopian green vision that provided much of the party's initial impetus and enthusiasm has been postponed – perhaps indefinitely. The mixed social-market industrial economy remains in place and – crucially – appears to have been accepted as such by a majority of the Greens. That this has happened was not just the inevitable outcome of pressures from the exogenous world, but also the result of decisions made within the party to accommodate to political reality.

Notes

1   Some scholars argue that the European elections provided an opportunity for the Greens to win a degree of state funding based on their performance (Kolinsky 1990: 67–8). However, it is not clear that the evidence supports this hypothesis, given that the Greens did not receive any state funding based on their performance in the 1979 elections until 1983 (Mintzel and Oberreuter 1992: 554). Nevertheless, given the Greens relatively small membership, the party has since become far more dependent upon state finance than the two main *Volksparteien.*

2 Marita Haibach was also appointed green adviser on women's issues, again demonstrating that the German Greens are a left-libertarian/new politics party rather than 'deep green'.
3 The NUKEM and ALKEM nuclear projects highlighted the problems of reconciling the new-left agenda with the 'old politics' of growth and production. Originally, the Greens had made the cancellation of funding for these projects the sine qua non of any co-operation with the SPD.

References

Bickerich, W. (ed.) (1985) *SPD und Grüne: Das neue Bündnis?*, Hamburg: Spiegel-Buch.

Broszat, M. (1981) *The Hitler State: the Foundation and Development of the Internal Structure of the Third Reich*, London, New York: Longman.

Bürklin, W. and Roth, D. (1993) *Das Superwahljahr: Deutschland vor unkalkulierbaren Regierungsmehrheiten*, Köln: Bund Verlag.

Dalton, R. J., Flangan, S. C. and Beck, P. A. (eds) (1984) *Electoral Change in Advanced Industrial Democracies: Realignment or Dealignment?*, Princeton, NJ, Guildford: Princeton University Press.

Döring, H. and Smith, G. (eds) (1982) *Party Government and Political Culture in Western Germany*, London, Basingstoke: Macmillan.

Dyson, K. (1982) 'Party government and party State', in H. Döring and G. Smith (eds), *Party Government and Political Culture in Western Germany*, London, Basingstoke: Macmillan.

Eckersley, R. (1989) 'Green politics and the new class: selfishness or virtue?', *Political Studies* 37, 2: 205–23.

Grant, W., Paterson, W. and Whitson, C. (1988) *Government and the Chemical Industry: a Comparative Study of Britain and West Germany*, Oxford: Clarendon Press.

Green, S. (1995) 'Postscript: the Land elections in Hessen, Nordrhein-Westfalen and Bremen', *German Politics* 4, 2: 152–8.

Harmel, R. and Janda, K. (1994) 'An integrated theory of party goals and party change', *Journal of Theoretical Politics* 6, 3: 259–87.

Hülsberg, W. (1988) *The German Greens: a Social and Political Profile*, London, New York: Verso.

Inglehart, R. (1990) *Culture Shift In Advanced Industrial Society*, Princeton, NJ: Princeton University Press.

Janda, K., Harmel, R., Edens, C. and Goff, P. (1995) 'Changes in party identity: evidence from party manifestos', *Party Politics* 1, 2: 171–96.

Katz, R. and Mair, P. (eds) (1992) *Party Organisations: a Data Handbook on Party Organisations in Western Democracies, 1960–90*, London: Sage.

Katzenstein, P. J. (1987) *Policy and Politics in West Germany: the Growth of a Semisovereign State*, Philadelphia, PA: Temple University Press.

Kitschelt, H. (1986) 'Political opportunity structures and political protest: anti-nuclear movements in four democracies', *British Journal of Political Science* 16, 1: 57–85.

Kleinert, H. (1992) Aufstieg und Fall der Grünen: Analyse einer alternativen Partei, Bonn: Verlag J.H.W. Dietz Nachf.

Kolinsky, E. (1990) 'The Federal Republic of Germany', in J. Lodge (ed.), *The 1989 Election of the European Parliament*, Basingstoke, London: Macmillan.

Lippman, W. (1914) *A Preface To Politics*, London: Mitchel Kennerley.

Markovits, A. S. and Gorski, P. S. (1993) *The German Left: Red, Green and Beyond*, Cambridge: Polity Press.

Mettke, J. R. (ed.) (1982) *Die Grünen: Regierungspartner von Morgen?* Hamburg: Spiegel–Buch.

Michels, R. (1970) Zur Soziologie des Parteiwesens in der modernen Demokratie, Stuttgart: Kohlhammer.

Mintzel, A. and Oberreuter, H. (1992) *Parteien in der Bundesrepublik Deutschsland, 2nd edn.*, Bonn: Bundeszentrale für Politische Bildung.

Müller, E.-P. (1984) *Die Grünen und das Parteiensystem*, Berlin: Deutscher Instituts–Verlag.

Padgett, S. (ed.) (1993) *Parties and Party Systems in the New Germany*, Aldershot, Brookfield, VT: Dartmouth.

—— (1994a) 'The German social democrats. A redefinition of social democracy or Bad Godesberg Mark II', in R. Gillespie and W. E. Paterson (eds), *Rethinking Social Democracy in Europe*, London, Portland, OR: Frank Cass.

—— (1994b) 'The German Social Democratic Party: between old and New Left', in D. S. Bell and E. Shaw (eds), *Conflict and Cohesion in Western European Social Democratic Parties*, London: Pinter.

Panebianco, A. (1988) *Political Parties: Organisation and Power*, Cambridge: Cambridge University Press.

Papadakis, E. (1983) 'The Green party in contemporary West German politics', *Political Quarterly* 54, 3: 302–7.

Paterson, W. E. (1982) 'Problems of party government in West Germany – a British perspective', in H. Döring and G. Smith (eds), *Party Government and Political Culture in Western Germany*, London, Basingstoke: Macmillan.

—— (1989) 'Environmental politics', in G. Smith, W. E. Paterson and P. H. Merkl (eds), *Developments in West German Politics*, Basingstoke, London: Macmillan.

Paterson, W. E. and Southern, D. (1991) *Governing Germany*, Oxford, Cambridge, MA: Blackwell.

Roberts, G. (1989) 'Political parties and public policy', in S. Bulmer (ed.), *The Changing Agenda of West German Public Policy*, Aldershot, Brookfield, VT: Dartmouth.

Scharf, T. (1994) *The German Greens: Challenging the Consensus*, Providence, RI, Oxford: Berg.

Smith, G., Paterson, W. E., Merkl, P. H. and Padgett, S. (eds) (1992) *Developments in German Politics*, Basingstoke, London: Macmillan.

van Deth, J. W. and Geurts, P. A. T. M. (1989) 'Value orientation, left-right placement and voting', *European Journal of Political Research* 17, 1: 17–34.

Weale, A. (1992) *The New Politics of Pollution*, Manchester, New York: Manchester University Press.

Primary data sources (Berlin and Hessen)

Abgeordnetenhaus von Berlin. Plenarprotokollen. Band I. 1989. 1 bis 18 Sitzung, Berlin 1991.

Ageordnetenhaus von Berlin. Plenarprotokollen. Band II. 1989/1990. 19 bis 31 Sitzung, Berlin, 1991.

Ageordnetenhaus von Berlin. Plenarprotokollen. Band III. 1990. 32 bis 50 Sitzung, Berlin, 1991.

Berliner Koalitionsvereinbarung zwischen SPD und AL vom 13. März 1989, SPD Ländesverband, 1989.

Berliner Statistik. Statistische Berichte. Wahlen in Berlin (West) am 29. Januar 1989. Endgültiges Ergebnis der Wahlen zum Abgeordnetenhaus und zu den Bezirksverordnetenversammlungen, Berlin: Statistisches Landesamt, 1989.

Berlin ist Freiheit. Eine starke SPD für Berlin. Wahlprogramme der Berliner SPD für die Wahlen zum Abgeordnetenhaus von Berlin am 29. Januar 1989. Beschlossen auf dem Landesparteitag am 15. Oktober 1988, SPD Ländesverband, 1988.

Das Kurzprogramme der Aternativen Liste. Die Kandidatinnen und Kandidaten für das Abgeordnetenhaus, Berlin: Alternative Liste, 1989.

Der Spiegel magazine.

Der Tagesspiegel.

Die Tageszeitung.

Handelsblatt.

Koalitionsvereinbarung für die 14. Wahlperiode des Hessichen Landtags zwischen Bündnis 90/ Die Grünen & SPD 1995–1999, Bündnis 90/Die Grünen, 1995.

Landtagswahlprogramm, Hessen '95, Bündnis 90/Die Grünen, 1995.

Rechenschaftsbericht der Fraktion Grüne/Alternativen Liste Berlin 1989/1990, Berlin: Alternative Liste, 1991.

Süddeutsche Zeitung.

9 Ecological modernisation, ecological modernities[1]

Peter Christoff

Ecological modernisation is emerging as a fashionable new term to describe recent changes in environmental policy and politics.[2] Its growing popularity derives in part from the suggestive power of its combined appeal to notions of development and modernity and to ecological critique. Yet competing definitions blur its usefulness as a concept. Does ecological modernisation refer to environmentally sensitive technological change? Does it, more broadly, define a style of policy discourse which serves to either foster better environmental management or to manage dissent and legitimate ongoing environmental destruction? Does it, instead, denote a new belief system or systemic change? Indeed, can it encompass all of these understandings? In this chapter, I want to examine current uses of the term in relation to the tensions between modernity and ecology, which it evokes, and suggest ways of diminishing its ambiguity.

It is widely acknowledged that, since the late 1980s, significant changes have occurred in the content and style of environmental policy in most industrialised (particularly OECD) countries. The nature and extent of these changes vary between nations,[3] reflecting their distinctive political institutional and cultural features; the national economic importance of the sectors and industries targeted by new regulatory regimes, and the extent and intensity of the environmental impact of those industries; the strength of popular environmental concern and of its political representation; the extent to which an 'implementation deficit' (the failure to realise environmental standards and goals) exists and is recognised as a local problem, and the reasons for this deficit; and regionally distinct perceptions of the key international and global ecological problems which mobilised public concern during the 1980s.[4]

Nevertheless, despite local variations, these environmental policy changes have several generalisable features. They have aimed to shift industry beyond reactive end-of-pipe approaches towards anticipatory and precautionary solutions which minimise waste and pollution through increasingly efficient resource use (including through recycling). Problem displacement across media (air, water, etc.) and across space and time has tended to be challenged by a more integrated regulatory approach – as much to achieve greater administrative efficiency and to limit regulatory overload as to address the new environmental problems caused by such displacement. Prescriptive regulatory approaches and 'technological forcing' – applied in the 1970s as the sole or predominant strategy for achieving ongoing improvements in environmental conditions – are more often accompanied or displaced by co-operative, voluntaristic arrangements between government and industry: increasingly, environmental-protection agencies seek to use industry's existing investment patterns and its capacity and need for technological innovation to facilitate improvement in environmental outcomes. A range of market-based environmental instruments have been deployed in response to the perceived exhaustion of the initial wave of regulatory intervention (Eckersley 1995). In all, the new environmental policy discourse increasingly emphasises the mutually reinforcing environmental and economic benefits of increased resource efficiency and waste minimisation.

These developments reflect an evolving international discourse in response to commonly perceived environmental problems. However, they also reflect an increasingly sophisticated political response by governments and industry to popular mobilisation around issues such as nuclear power, acid rain, biodiversity preservation, ozone depletion and induced climate change. In other words, the new policy culture and its trends are not always simply or primarily intended to resolve environmental problems. They are also shaped by a contest over political control of the environmental agenda and, separately, over the legitimacy of state regulation (predominantly in the English speaking OECD countries). In addition, they have been influenced by the growing pressures on nation states generated by intensified economic globalisation and by changes in the structure and nature of production towards greater flexibility and international integration.

The strengthening of linkages between environmental and economic policy is especially observable in countries such as Germany – in turn raising questions about the reasons for their exceptionally good environmental performance in contrast to countries such as the UK and the US. For instance, in the 1980s and the early 1990s German environmental policy, under pressure from the Greens, has moved rapidly to address its failure to meet targets and standards adopted during the 1970s. The promotion of design criteria enabling comprehensive reuse of materials has been accompanied by regulations requiring that 72 per cent of glass and metals and 64 per cent of paper board, laminates and plastics be recycled by 1995 (Moore 1992). Regulations also encourage use of 'waste energy' for heating and power generation. The 1983 Large Combustion Plant Ordinance requires the retrofitting of all major power plants to cut pollutants contributing to acid rain by 90 per cent by 1995. Laws passed in 1989 ban chlorofluorocarbon (CFC) production and use by 1995. Germany has also committed itself to a unilateral reduction in carbon-dioxide emissions of between 25 and 30 per cent by the year 2005 and, since 1990, has begun to articulate and implement a package of some sixty measures to enable it to meet this target (Hatch 1995).

These changes have been supported by considerable government assistance. Weale (1992) reports that between 1979 and 1985, the German government subsidy for environmental research and development rose from $US 144.3 million to $US 236.4 million, or from 2.1 per cent of research and development to 3.1 per cent (the UK equivalent was a 0.8 per cent rise to 1.1 per cent). This commitment is also institutionally defined: Germany has a separate federal Ministry for Research and Technology which spends about DM 200 million per annum on research and development of environmental technologies (Angerer 1992: 181). Since 1985, the level of German public subsidy for environmental research has exceeded that of the US in absolute terms. Public investment also provides substantial support in the energy-conservation fields, including research into energy-conservation devices. As a result, within a decade German industry has become a global leader in the development and/or production of solar photovoltaics, high-efficiency turbines, hydrogen-powered cars, energy-efficient household appliances and recyclable materials and products.

The economic advantages to countries and companies leading the field in environmental performance improvements have been recognised as considerable. It is estimated that by the year 2000, Japan will be producing some $US 12 billion worth of waste incinerators, air-pollution equipment and water-treatment devices, and the Ministry of International Trade and Industry (MITI) has proposed aid projects aimed at energy development in China, Indonesia and Malaysia as a means of further tying and strengthening trade connections with these countries (Gross 1992).

The extent of formal policy integration and of the diffusion of environmental principles into the practices of state, economy and society is also of particular interest. Governments in several countries – notably Australia, Canada, the Netherlands and the UK – have developed national plans for sustainable development: metapolicies aimed at the integration of national environmental and economic activity and at encouraging greater environmental awareness in civil society. In this, the Dutch National Environment Policy Plan (NEPP) has been significantly more successful than similar attempts elsewhere. This success is partly due to the highly corporatist nature of Dutch politics and planning; the Dutch state's acceptance of a significant role in facilitating and directing industrial development and environmental protection; and also the timing of the plan's release in 1989, during a high point of international and national environmental concern. The NEPP has the explicit goal of achieving environmental sustainability in the Netherlands within one generation, by 2010, by recasting policies and practices in key economic sectors – including manufacturing, agriculture and transport – to limit waste production and environmental pollution (Carley and Christie 1992; van der Straaten 1992). Despite weaknesses both in its targets and ongoing implementation (Wintle and Reeve 1994), the plan nevertheless offers a programmatic approach to working towards measurable targets against which the public, government and industry can assess its progress and iteratively adjust the Plan.[5]

Over the past decade, Germany, the Netherlands, Japan and the Scandinavian block appear to have achieved above OECD-average improvements across a range of industry related national environmental indicators, including water quality and air-pollutant emissions (OECD 1993). In these countries there is now evidence of a decoupling of gross national product (GNP) growth from the growth of environmentally harmful effects, indicating increased economic output with decreased energy and materials consumption *per unit* of GNP. However, certain improvements in environmental conditions in the First World have been gained through displacement of high-energy-consuming and/ or polluting industries (e.g. metal processing and primary manufacturing) to newly industrialising countries (NICs) and lesser-developed countries (LDCs). Meanwhile, underlying increases in total material consumption in both industrialising and industrialised countries continue to enhance environmental pressures, suggesting both that the pace of reform is too slow and the root cause of the 'implementation deficit' of the 1970s has not been overcome (WRI 1994).

The uses of ecological modernisation

Positive aspects of these recent changes have been described by academic observers as evidence of a process of ecological modernisation, although their uses of the term vary considerably in scope and meaning. Specifically, leading exponents of the term in the German and English literature – such as Jänicke, Hajer and Weale – use it in their policy analysis, sociological analysis or political-theoretical discussion in ways which are occasionally problematic, partly because of a lack of clarity about whether the term is being used descriptively, analytically or normatively. Following discussion of these distinctive uses of ecological modernisation, I want to propose a typology for ecological modernisation which emphasises the normative dimensions of the term.

Ecological modernisation as technological adjustment

Ecological modernisation has been used narrowly to describe technological developments with environmentally beneficial outcomes – such as chlorine-free bleaching of pulp for paper and more fuel-efficient cars – specifically aimed at reducing emissions at source and fostering greater resource efficiency (Simonis 1988; Jänicke 1988; Zimmermann *et al.* 1990; Jänicke *et al.* 1992).

Jänicke, who perhaps first introduced ecological modernisation into the language of policy analysis (Jänicke, 1990), for instance, refers to four broadly framed 'environmental political' strategies commonly found in industrial countries (Jänicke 1988). Two of these strategies are remedial (compensation and environmental restoration, and technical pollution control) and two preventative or anticipatory (environment-friendly technical innovation or ecological modernisation and structural change). For Jänicke, ecological modernisation is fundamentally a technical cost-minimisation strategy for industry and an alternative to labour-saving investment – a form of 'ecological rationalisation' which will lead simultaneously to greater 'ecological and economic efficiency' (Jänicke 1988: 23). It is primarily seen as a strategy intended to maintain or improve market competitiveness, in which the environmental benefits of such technological change are incidental rather than a core concern for innovation and implementation. In this sense, such a narrow version of ecological modernisation does not necessarily reflect any significant and overwhelming changes in corporate, public or political values in relation to desired ecological outcomes. Rather, it is an outcome of capital's cost-minimising responses to new pressures – such as the adoption elsewhere of post-Fordist 'lean' production methods (Best 1990; Amin 1994; Wallace 1995), resource price movements and scarcities (e.g. the oil crises of the 1970s), changes in consumer taste, and profit squeezes caused by taxes and regulatory strategies of the state – at a time when automation has reduced industry's capacity to increase labour's productivity. Innovation and implementation may be confined to those areas and types of technical improvements which ensure market competitiveness.[6] Consequently, such technological change may not contribute to lasting environmental improvements when viewed in the context of national or international ecological requirements.

For Jänicke, moves towards sustainability depend on broader structural change – the second of his anticipatory strategies – which would lead to profound shifts in production and consumption patterns. These are not merely industrial responses to ecological symptoms (e.g. resource shortages), but incorporate precautionary analysis and associated restrictions on action, and lead to constrained 'qualitative' economic growth and a decrease in absolute resource use, pollution and environmental degradation (Jänicke 1988: 15–17, 1992). He sees the current period of multiple crises – unemployment, accumulation, environmental degradation and state finance (fiscal crisis) – which extends into the 1990s as also providing opportunities for the 'creative destruction' of old patterns and forms. The world market involves not only competition between enterprises producing new technologies but also competition between nations with stronger and weaker 'state steering capacity', a competition favouring those capable of breaking with the tendency to protect their old 'smoke-stack industries' and able to generate a framework for consensual transformation. Jänicke's more recent empirical work documents the sites, conditions and (limited) signs of such industrial transformation (Jänicke 1992).

Jänicke and his colleagues fail to identify or address potential political economic contradictions in this narrow vision of an ecological modernisation embedded in larger processes of structural transformation. At what point are the currently developing patterns of unrestricted, globalised production and trade, and the cultural demands for increasingly specialised consumption, challenged? How will the corresponding growth in international markets for new 'lean' technologies and products be restrained to ensure regional and international ecological stability, rather than ongoing expansion of total resource use and waste output? And, specifically given Jänicke's views (Jänicke 1992) on state failure and the limits of state action, what institutions will participate in this enhanced process of regulation? What happens, within this larger scenario, to those countries – the technological laggards – unable to compete or perform economically and ecologically?

If these new, clean technologies and products *are* truly ecologically sustainable – leading to a significant absolute decrease in resource use and to effective environmental preservation – what are their ramifications for trade, employment, accumulation and wealth distribution within and between nations, particularly if they are sought according to time frames which are dictated by urgent ecological demands (e.g. the potential need to cut greenhouse-gas emissions by up to 60 per cent within the next three decades)? Certainly, given its narrowly industrial focus, such ecological modernisation would not necessarily serve to diminish total resource consumption or lead to the protection of 'unvalued', non-resource-related ecological concerns.

Ecological modernisation as policy discourse

Others, such as Weale (1992, 1993) and Hajer (1995), have employed ecological modernisation more broadly to define changes in environmental policy discourse. For Hajer, the shift toward ecological modernisation can be observed in at least six 'realms' namely: in environmental policy-making, where anticipatory replace reactive regulatory formulae; in a new 'proactive' and critical role for science in environmental policy-making; at the microeconomic level, in the shift from the notion that environmental protection increases cost to the notion that 'pollution prevention pays'; at the macroeconomic level, in the reconceptualisation of nature as a public good and resource rather than a free good; in the 'legislative discourse in environmental politics', where changing perceptions of the 'value' of nature mean that the burden of proof now rests with those accused as polluters rather than the damaged party; and the reconsideration of participation in policy-making practices – with the acknowledgement of new actors, 'in particular environmental organisations and to a lesser extent local residents' – and the creation of 'new participatory practices' for their inclusion in a move to end the 'sharp antagonistic debate between the state and the environment movement' (Hager 1995: 28–9).

Hajer predominantly regards ecological modernisation as a policy discourse which assumed prominence around the time of the European Community's Third Action Plan for the Environment and, more explicitly still, the 1984 OECD Conference on Environment and Economics. Such ecological modernisation 'recognises the structural character of the environmental problematique but none the less assumes that existing political, economic and social institutions can internalise care for the environment' (Hajer 1995: 31). Hajer (1995: 101) sees this discourse as being largely economistic: framing environmental problems in monetary terms, portraying environmental protection as a 'positive sum' game, and following a utilitarian logic. At the core of ecological modernisation is the idea that pollution prevention pays: it is 'essentially an efficiency-oriented approach to the environment'. In other words, economic growth and the resolution of environmental problems can, in principle, be reconciled (Hajer 1995: 25–6).[7] 'Ecological modernisation uses the language of business and conceptualises environmental pollution as a matter of inefficiency while operating within the bounds of cost effectiveness and bureaucratic efficiency' (Hajer 1995: 31).

Hajer is most effective where he suggests that ecological modernisation is a discursive strategy useful to governments seeking to manage ecological dissent and to relegitimise their social regulatory role. It permits a critical distancing from the interventionist remedies of the 1970s which, as Hajer believes, 'did not produce satisfactory results' and may serve to legitimate moves to roll back the state and reduce its regulatory capacities in the environmental domain. It also enables governments to promote environmental protection as being economically responsible, thereby resolving the tensions created by previous perceptions that the state was acting against the logic of capital and its own interests (of functional dependency on private economic activity). He suggests that such a strategy explicitly avoids addressing basic social contradictions that other discourses might have introduced.

> Ecological modernisation is basically a modernist and technocratic approach to the environment that suggests that there is a techno-institutional fix for present problems. Indeed ecological modernisation is based on many of the some institutional principles that were already discussed in the early 1970s: efficiency, technological innovation, techno-scientific management, procedural integration and co-ordinated management. It is also obvious that ecological modernisation as described above does not address the systemic features of capitalism that make the system inherently wasteful and unmanageable.
>
> (Hajer 1995: 32)

In other words, ecological modernisation is not simply a technical answer to the problem of environmental degradation. It can also be seen as a strategy of political accommodation of the radical environmentalist critique of the 1970s; meshing with the deregulatory moves which typify the 1980s; with distinctive affinities with the neoliberal ideas that dominated governments during this time; supporting their concern for structural industrial reform (Hajer 1995: 32–3).

Hajer's own views here are unclear. He seems to approve of such political closure yet leaves open the question of whether or not ecological modernisation might be, in the terms of the critics of Brundtland, a rhetorical ploy to take the wind out of the sails of 'real' environmentalists, one which displaces and marginalises the radical emancipatory aspects of environmental critique (Hajer 1995: 34). He is even less clear about whether or not ecological modernisation 'may not in fact have a more profound meaning [...] as the first step on a bridge that leads to a new sort of sustainable society'. Hajer mainly sees ecological modernisation as a counter to the 'antimodern' sentiments he claims are part of the critical discourse of new social movements.

> It is a policy strategy that is based on a fundamental belief in progress and the problem-solving capacity of modern techniques and skills of social engineering. Contrary to the radical environment movement that put the issue on the agenda in the 1970s, environmental degradation is no longer an anomaly of modernity. There is a renewed belief in the possibility of mastery and

control, drawing on modernist policy instruments such as expert systems and science.

(Hajer 1995: 33)

In this sense too, as it seeks to provide a soothing rhetoric promoting apparent remedial and anticipatory change, such a policy discourse may be profoundly anti-ecological in its outcomes, its narrow economism serving to devalue and work against recognition and protection of non-materialistic views of nature's 'worth'.

Ecological modernisation as belief system

Both Hajer and Weale also use the concept in more radical ways.[8] For Weale (1992), ecological modernisation represents a new belief system that explicitly articulates and organises ideas of ecological emancipation, which may remain confused and contradictory in a less self-conscious discourse. It is an ideology based around, but extending beyond the understanding that environmental protection is a precondition of long-term economic development. Weale's claims for ecological modernisation as a belief system are important given the role of belief systems in organising and legitimising public policy.

Weale also sees ecological modernisation as being focussed on a reconceptualised relationship between environmental regulation and economic growth. It still includes an emphasis on achievement of highest possible environmental standards as a means for developing market advantage through the integration of anticipatory mechanisms into the production process; recognition of the actual and anticipated costs of environmental externalities in economic planning; and the economic importance of strengthening consumer preferences for 'clean, green products' (1992: Chapter 3, 1993: 206–9). However, 'once the conventional wisdom of the relationship between the environment and economy is challenged, other elements of the implicit belief system [which sees them in opposition] might also begin to unravel'. Regulation may 'no longer seem merely a mechanical matter'. Ecological modernisation thus prefigures systemic change and may, in its more radical forms, generate a broader transformation in social relations, one which leads to the ecologisation of markets and the state.

Under such circumstances, as Weale comments,

the internalisation of externalities becomes a matter of attitude as well as finance, and a cleavage begins to open up not between business and environmentalists, but between progressive, environmentally-aware business on the one hand and short-term profit takers on the other. Moreover, the behaviour of consumers becomes important, so that the role of government policy is not simply to respond to the existing wants and preferences of their citizens, but also to provide support and encouragement for forms of environmentally aware behaviour and discouragement for behaviour that threatens or damages the environment. Once this view has taken root, the line from mechanical to moral reform has been crossed. The challenge of ecological modernisation extends therefore beyond the economic point that a sound environment is a necessary condition for long-term prosperity and it comes to embrace changes in the relationship between the state, its citizens and private corporations, as well as in the relationship between states.

(Weale 1992: 31–2)

However, in *The New Politics of Pollution,* Weale (1992) does not develop his views on the transformations of both civil society and the state necessary to achieve ecological sustainability. What limits are posed by the state's dependent relationship to private sector economic activities? How can these be overcome given the increasing political and economic vulnerability of individual nation states to global flows of capital? To what extent would transformations of civil society and in public spheres, rather than institutional changes to the state, drive the process of ecologisation?[9]

Ecological modernisation – some unresolved issues

It is possible to illuminate problems and issues left unaddressed or unresolved by the foregoing uses of ecological modernisation by asking a series of interrelated questions. In different situations (different policy forums and different countries) quite different styles of ecological modernisation may prevail – ones which can be judged normatively to tend towards either 'weak' or 'strong' outcomes on a range of issues, such as ecological protection and democratic participation. In this sense, these questions hint at the limitations of those forms of ecological modernisation which tend toward the first rather than the second of what might seem, initially, opposing 'poles'.

Economistic or ecological?

In each of the uses of ecological modernisation described above, the environment is reduced to a series of concerns about resource inputs, waste and pollutant emissions. As cultural needs and 'non-anthropocentric' values (such as are reflected in the Western interest in the preservation of wilderness) cannot be reduced to monetary terms, they tend to be marginalised or excluded from consideration. This is clearly the case for ecological modernisation narrowly defined as technical innovation. But it is equally true of those interpretations of ecological modernisation which see the state shaping corporate activity and markets to (re)incorporate environmental 'externalities' into the costs of production. As has been noted, such versions of ecological modernisation may remain consistent with the traditional imperatives of capital. Leading industries may welcome uniformly-applied environmental regulatory regimes, as the redefinition of the boundaries of acceptable economic behaviour may represent a rationalisation of their markets which makes the rules of production and competition more certain or amenable to their entry or dominance. But ideologically and practically, such ecological modernisation may simply put a green gloss on industrial development in much the same way that the term 'sustainable development' has been co-opted: to suggest that industrial activity and resource use should be allowed as long as environmental side effects are minimised.

Given this dominant emphasis on increasing the 'environmental efficiency' of industrial development and resource exploitation, such ecological modernisation remains only superficially or weakly *ecological*. Consideration of the integrity of ecosystems, and the cumulative impacts of industrialisation upon these, is limited and peripheral. In this sense the entire literature is somewhat Eurocentric: deeply marked by the experience of 'local' debates over the politics of acid rain and other outputs, rather than conflicts over biodiversity preservation. Although current uses of ecological modernisation may be well adapted to describing positive environmental outcomes in certain industrialised First World countries, where a version of ecological sustainability may be created in the wasteland of a vastly depleted biological world, it may be positively dangerous if taken prescriptively by those nations where the conservation of biodiversity is a more fundamental concern or opportunity and/or which depend on primary resource exploitation to fund their traditional forms of economic growth (e.g. Australia, Brazil and South Africa).

National or international?

The uses of ecological modernisation described earlier also remain narrowly focussed on changes within industrialised nation states. They are therefore unable to integrate an understanding of the transformative impact of globalisation on environmental relations at all political levels. They offer only a diminished recognition of the increasingly internationalised flows of material resources, manufactured components and goods, information and waste; of the influence of multinational corporations on investment, national industrial development and the regulatory capacities of the nation state; and of international deregulatory developments (such as GATT) and environmental treaties (such as the Montreal Protocol). Paradoxically, each of these facets of globalisation shapes yet distorts, provokes yet inhibits, and undermines the emergence of 'strong' forms of ecological modernisation at national and regional levels. Because of their nation statist focus, these uses of ecological modernisation – including those raising broader ideological and systemic concerns – still tend to remain focused on localised end-of-cycle issues rather than encompassing the globally integrated nature of resource extraction and manufacturing in relation to domestic consumption, overvaluing 'local' achievements and environmental impacts while undervaluing geographically distant factors.[10]

Consider, for instance, the internationally dispersed resources and environmental impacts associated with producing and running a nation's car fleet, or with producing and using paper; or the extent to which heavy transformative industries such as smelting, ship building or car manufacturing have relocated to the NICs. In other words, although primary consumption of energy and other primary resources may have fallen in relation to GNP in certain European econo-mies as these have become increasingly 'post-industrial', their *per capita* material consumption continues to grow – with their environmental impacts now displaced 'overseas'. In addition, reliance on the emergence of 'leading' post-industrial economies and the domination of global markets by a country with the most stringent pollution-control standards (Weale 1992: 77), perhaps misunder-stands the structure and operation of an increasingly globalised capitalism.

Given this presently, predominantly nation-statist view of ecological moder-nisation, discussion of the emergent international institutions for environmental regulation and protection and of environmental trends remains underdeveloped where it occurs in the ecological modernisation literature.[11] The literature fails to recognise that, because 'old' forms of industrial activity with their associated environmental problems are being displaced to 'developing' nations or regions and the transition to alternative technologies is occurring too slowly to prevent major global environmental problems (such as climate change), we may instead be moving towards what Everett (1992) has called the 'breakdown of technolo-gical escape routes', as the ecological pressure for change increases beyond the reasonable capacities for social and industrial reform.

Hegemonic progress or multiple possibilities?

In different ways, the types of ecological modernisation described earlier are also presented as contributing to, or constituting, a unilinear path to ecological moder-nity. Consequently they seem to be offering a revival of mainstream development theory and of notions of 'uneven development' and 'underdevelopment', positing ecological modernisation as the next necessary or even triumphant stage of an evolutionary process of industrial transformation – a stage dependent upon the hegemony of Western science, technology and consumer culture and propagated by leading Western(ised) countries with the appropriate intellectual and economic capacity. Such views of ecological modernisation may be validly subjected to the criticisms which were levelled against development theory two decades ago.

Theorists adopting this view, implicitly or explicitly, rely upon a simplistic division between 'traditional' and 'modern' societies and ignore the potential for a multiplicity of paths to ecological sustainability which may rest in the diversity of non-Western cultures. They seem to suggest that all countries may undertake the great leap forward over the phase of 'dirty' industrialisation into the fully ecologically modern, post-industrial condition. But if 'ecologically modernising countries' cannot quite manage the great leap, then at least such nations will eventually be able to employ restorative technologies, salves, and panaceas developed elsewhere to undo the ecological devastation resulting from the stage of aggressive industrialisation to which developing countries aspire or are now subject. In other words, when developing countries reach the levels of affluence which give them the economic capacity to afford 'ecological moderni-sation', they will be able to turn to consider and repair the path of devastation which has bought them this luxury. In fact, such views of ecological modernisation offer a world divided by renewed or strengthened 'core–periphery' relationships between industrialised and industrialising countries, with domination of world markets, and the 'motors' of progress dominated by leading industrial state(s).

The problems here are most obvious when we consider the potentially disastrous local and global ecological (and cultural) costs of China, India, Indonesia or Brazil pursuing such a path to ecological modernity, or the perpetual mendicant status of small nation states such as the Solomons or Vanuatu, and also much of the African continent once trapped in this cycle of cultural and technological dependency.

Technocratic or democratic?

There are also tensions between what different writers describe as the preconditions for systemic or structural ecological modernisation. Some stress the transformative impact of environmental awareness on civil society and the public sphere, and on the institutions and practices of government and industry. They emphasise the ways in which citizenship and democratic participation in planning may serve to socialise and 'ecologise' the market and guide and limit industrial production. Others, however, favour a technocratic, neocorporatist version of ecological modernisation – one which may prove primarily a rhetorical device seeking to manage radical 'antimodern' dissent and secure the legitimacy of existing policy while delivering limited, economically-acceptable environmental improvements.

For example, Weale (1992), interpreting developments in Germany and the Netherlands, suggests that the systemic realisation of ecological modernisation requires a proactive, interventionist state supporting a well-developed culture of environmental policy innovation and offering significant public investment and subsidies as means for achieving economic advantage and environmental outcomes. Such state activity would entail an integrated regulatory environment; strong structural and process cross-linkages between different parts of the state; and the development of a synoptic and reflexive use of environmental information in policy formation and implementation. In addition, this transformation is enhanced by, or indeed *depends upon*, increased public participation in political decision-making, including green political pressure through both the environment movement and parliamentary politics (including Green parties); and increased public influence over industry behaviour through 'green consumer' action and the activities of environmental pressure groups and organisations.

By contrast, Hajer (1995) and Andersen (1993) seem to believe that a more technocratic relationship between state and civil society leads more effectively to systemic ecological modernisation. Andersen (1993), who is specifically concerned to define preconditions for ecological modernisation through comparative analysis of national environmental performance, describes a country's *capacity* for ecological modernisation as depending upon its 'achieved level of institutional and technological problem solving capabilities, which are critical to achieving effective environmental protection and transformation to more sustainable structures of production'. He argues, in concert with Jänicke and others, that there is also a close relationship between consensus-seeking policy styles and high levels of environmental protection in industrialised countries.

Andersen suggests that four basic variables govern the capacity for such ecological modernisation:

1. *economic performance*, the capacity of countries to pay for environmental protection – a factor which appears directly linked to the intensity of environmental pollution;
2. *consensus ability*, which Andersen believes is best developed in countries with neocorporatist structures, which are seen as having consensus-seeking decision-making styles that are more amenable to dealing with new ideas and interests;[12]
3. *innovative capability*, which he describes as the capacity of both the state and the market institutions to remain open to new interests and innovations in the judicial and political system, the media and economic system; and
4. *strategic proficiency* – the capacity to institutionalise environmental policy across sectors. He identifies federal states, which face potential fragmentation and delay in implementation, and states evidencing strong compartmentalisation of the bureaucracy – concomitant with weak environment departments or agencies – as potentially suffering problems in this area (Andersen 1993: 3).

Andersen proposes that the presence of these variables or attributes seems to contribute to, or at least correlate with, the success of 'leading' European countries – such as Germany and the Scandinavian bloc – in achieving exceptional improvements in environmental conditions. But how do they apply to the NICs and LDCs? Again, what relationship between state and civil society and what forms of democratic participation are required, especially given the international dimensions of environmental problems, to enable the radical social and economic changes which ecological sustainability may require?

Insofar as ecological modernisation focuses on the state and industry, in terms which are narrowly technocratic and instrumental, rather than on social processes, in ways which are broadly integrative, communicative and deliberative, it is less likely to lead to the sorts of embedded cultural transformation which could sustain the major demands of reduction of material consumption levels, significant and rapid structural transformations in industrialised countries, and major international redistributions of wealth and technological capacity. In general, the extent and nature of institutional changes required to enable the full recognition of a discursive and participatory environmental politics (and accommodate the transboundary and intertemporal nature of environmental risks and impacts) have not yet been explored in the ecological modernisation literature.[13]

Table 9.1 Types of ecological modernisation

Weak	Strong
Technological (narrow)	Institutional/systemic (broad)
Instrumental	Communicative
Technocratic/neocorporatist/closed	Deliberative democratic/open
National	International
Unitary (hegemonic)	Diversifying

Weak and strong ecological modernisations

Given the range of uses to which the term has been put, can ecological modernisation be rescued and 'stabilised' as a concept? There is a need to differentiate between sometimes conflicting versions of ecological modernisation. These versions do not each merely describe some aspect of a more encompassing process of ecological modernisation but offer quite different real-world outcomes. Some of these uses may be labelled 'narrow' or 'broad', depending on the extent to which they are technological or systemic in scope or focus. More importantly, and reflecting the above discussion, it is instructive to draw out the normative dimensions of different versions of ecological modernisation. I would suggest that different interpretations of what constitutes ecological modernisation lie along a continuum from 'weak' (one is tempted to write 'false') to 'strong', according to their likely efficacy in promoting enduring ecologically sustainable transformations and outcomes across a range of issues and institutions (Table 9.1). The political contest between the environment movement and governments and industry is predominantly over which of these types of ecological modernisation is to predominate.

It is essential to note that 'weak' and 'strong' features of ecological modernisation are not always mutually exclusive binary opposites. Some features of 'weak' or 'narrow' ecological modernisation are necessary but not sufficient preconditions for an enduring ecologically sustainable outcome. Clearly, one does not abandon technological change, economic instruments or instrumental reason in favour of institutional and systemic change or communicative rationality. In many cases – although not all (e.g. technocratic or neocorporatist versus deliberative and open democratic systems) – aspects of 'narrow' or 'weak' ecological modernisation need to be subsumed into and guided by the normative dimensions of 'strong' ecological modernisation.

Ecological modernisation, ecological modernities

The uses of ecological modernisation discussed above do not address tensions and contradictions embedded in the term at the point where ecological critique[14] challenges the ways in which simple industrial modernisation defines its relationship to nature. Perhaps the most radical use of ecological modernisation would involve its deployment against industrial modernisation itself. To understand what this might mean, it is necessary to unpack the ecological and modernising components of ecological modernisation and look at their interaction more closely.[15]

Modernity has broken or swept aside traditional forms of order and certainty: as Marx put it, 'all that is solid melts into air'. Its dynamism may be attributed to the separation of time and space into a realm which is detached from immediate experience, the disembedding of social systems and the reflexive ordering and reordering of social relations (Giddens 1990: 16–17). By fostering relations with 'absent' others, locationally distant from any given situation of face-to-face inter-action, the process of modernisation increasingly overlays place (the immediate experience of location) with space (the abstract experience of location, into which the immediate experience of location is then fitted). It also replaces 'local time', based on an immediate experience of the rhythms of Nature and the require-ments of one's immediate community, with 'abstract time' – now most powerfully represented by the international acceptance of a standard differentiation of global time zones. The extreme dynamism of modernity, Giddens argues, also depends on the establishment of *disembedding* social institutions, which create or support the creation of abstract social relations and their associated organisations. The emergence of symbolic tokens (such as money) and expert systems represents an essential feature of modernity and contributes centrally to this process of disem-bedding, which is then reflected, for instance, in increasingly global discourses such as in science or law.

In addition, 'systems of technical accomplishment or professional expertise organise large areas of the material and social environments in which we live today' (Giddens 1990: 27). Crucial among these systems are those of scientific understanding and technological performance. We live in and are dependent upon – that is to say, *trust* in – them for our survival and legitimate functioning. These abstract expert systems constitute not only bodies of knowledge, but also lived forms of social relationship. We enter them whenever we turn on a light switch, step on an aeroplane, go to the dentist or answer the telephone: their complexity and functioning are taken for granted in a socially learned, (relatively) unquestioned and automatic way.[16] Both types of disembedding mechanism – symbolic tokens and abstract expert systems – provide 'guarantees' of expecta-tions across time and space and 'stretch' social systems as a result. They also promote a new awareness of risk, which is the product of the human-created technological and social characteristics of modernity. Risk and trust intertwine. Modernity is also notable for the development of new capacities for the reflexive appropriation of knowledge, in part born of the capacity to transmit and review, which comes with the development of the book and other forms of recorded information. All knowledge and beliefs become available for scrutiny. Certainty is displaced.

These features together contribute to the emergence of the institutional dimen-sions of modernity. Giddens (1990: 59) identifies four such dimensions, which are interrelated and interdependent: capital accumulation, industrialism, surveil-lance and military power. Of particular interest in relation to ecological modernisation are the first three dimensions. Industrialism seeks the transformation of nature into created or recreated (managed) environments through processes of standardisation, rationalisation and reduction. The imperatives of capital accumulation are such that the hunt for markets and resources encourages the commodification of all aspects of individual cultures and nature which remain vulnerable. The capacity for surveillance – in the broadest sense, in terms of the apparatuses of consolidated administration, monitoring and registering of social and environmental facts – has a bearing on the development of modern forms of reflexive environmental management.

The last point to note here is the globalising tendency of modernity. Its global reach is partly a result of the imperial and colonising tendencies of capital accumulation. However, as modern technologies of transport, information transfer and communication continue actively to redefine social relations, linking and integrating distant parts of the globe both as markets for commodities and abstract social networks, the notions of 'centre' and 'periphery' begin to blur. The flow of individuals, commodities, cultures and pollution across territorial borders is also leading to a practical redefinition of one of the other major institutions of modernity: the nation state. Modernity brings with it the globalisation of risk by altering the scope, the type and the range of human-created environmental risks which individuals now face, and also the globalisation of the perception of these risks.

How then can we characterise the relationship between 'modernisation' and 'the ecological'? Modernity is fraught with tensions and generates its own new contradictions: nowhere is this more evident than in relation to the environment. The emergent ecological critique of untrammelled industrialism – sharpened politically in recent years by perceptions of ecological crisis and of the need for precautionary consideration of the potential consequences of development – has a paradoxical relationship to the constitutive features of modernity described above. Itself a product of simple modernity, ecological critique both depends upon and resists the modern reorganisation of time and space. It makes radically problematic and contradictory the industrialising imperative which lies at the heart of modernisation by redefining the cultural and ecological limits to the instrumental domination of Nature.

Ecologically re-embedding space and time

The 'birth' of nature has been accompanied and shaped by the simultaneous creation of technological forces which lead to what McKibben (1990) has called 'the death of nature' through human interference with previously autonomous natural systems world-wide (through induced climate change, the global transport of pollutants and so on). Driven by the imperatives of capital accumulation, industrialism – shaped by the alliance of science and technology – continues to transform nature in ways unimaginable to earlier generations. It does so deliberately, for instance, by introducing alien plant and animal species to new continents, by flooding valleys and levelling mountains, and by creating new relations of physical and economic dependency between the country and the city, and between the First and the Third Worlds. Colonial conquests have often also led to the unintended extermination of indigenous plants and animals through destruction of their native habitat or by introduced predators. Demand for export earnings and the development of industrialised, monocultural agriculture and forestry have produced a wave of extinction that continues to roll across North and South America, Australia, Asia and Africa. Yet the ecological transformation threatened by the combined impacts of induced global warming and biotechnology is more comprehensive still.

The creation of a secular, scientific understanding of nature – indeed, the development of ecology as a scientific discipline – and the 'triumph' of technological domination over natural cycles and ecological processes, depend upon and arise from the separation of time and space discussed earlier. The 'discovery' of 'remote' regions and 'exotic' species enabled the scientific conceptualisation of natural systems at the same time as it involved the commodification of those environments and the imperial domination or appropriation of non-Western knowledge of natural systems and species.

Yet, the 'so-called economy of nature, the interrelationship of all organisms' for which Haeckel (1870) coined the term 'ecology' in 1869, depends on cycles and time scales which are generally alien to those of the political and economic institutions of industrial society. An ecological critique recognises and respects the importance of the cycles upon which the biological world depends, and which seeks to re-embed our relationship to Nature in a local place and to redefine the relationship in ecological-temporal terms, often stands in opposition to the transcendent, abstracting features of modernity (and its industrial manifestations) while still to some extent depending upon its conceptual frameworks. In other words, such ecological critique tries to undo the stretching of time and space as it seeks to limit certain aspects of industrial modernisation in order to preserve the ecological integrity of natural systems or to preserve cultural understandings and institutions which are locally embedded and resistant to the resource utilitarianism of all forms of industrial modernity.[17]

Let me give several examples of such critique, each relating to primary resource use. Harvesting temperate forests on an *ecologically* sustainable cycle which also respects the needs of dependent species may involve 300- or 400-year rotations and it is probably ecologically out of the question for complex, fragile rainforest systems. As such time spans may be commercially unviable, protection of non-resource species involves fundamentally rethinking how or whether one can use these forests. The international ban on whaling, based on moral considerations, defies the industrial/instrumental belief in the potential for whales to be 'harvested' sustainably. For similar reasons, environmentalists now campaign to preserve wilderness areas and for animal rights. Also consider the conflicts between the environment movement and industry over the *representations* of place versus space; struggles with profound material consequences for particular rivers, forests and wetlands. One may recognise a fierce contest over images and counter-images of sites in the diametrically opposed terms used by developers and environmentalists to represent contested terrain through the Australian media as either significant places or exploitable spaces with few speci-fically valuable attributes – 'the last free river' versus a 'leech-ridden ditch' for the Franklin River in Tasmania; 'magnificent Northern wilderness' and 'sacred ground' versus 'clapped-out buffalo country' for Coronation Hill, a proposed mine site in the Northern Territory. As Harvey (1993: 23) notes, in such cases 'the cultural politics of places, the political economy of their development, and the accumulation of a sense of social power in place frequently fuse in indistin-guishable ways'.

Each of these examples stresses ways in which a non-economistic ecological critique is in tension with or begins to break away from industrial modernity, even as it still uses the media, scientific information and political institutions, which are products of late modernity as its tools. In other words, ecological critique is not (as Hajer would suggest) naively antimodern, seeking to dismantle all abstract relations established through modernisation. Rather, as the product of modernity and something which continues to depend upon modernity's processes for its development,[18] it aims to discipline and restrain – to put bind-ings, brakes and shackles (Offe 1992) – on the overdetermining effects of globa-lised productive systems. Beck (1992: 23) writes of how the life of a blade of grass in the Bavarian forest ultimately comes to depend on the making and keeping of international agreements: ecological critique requires abstract systems and their institutions and ecological considerations to coexist through the prioritisation of the latter.

The strongest or most radically *ecological* notion of ecological modernisation will often stand in opposition to industrial modernity's predominantly instrumen-tal relationship to nature as exploitable resource. Recognition that overproduc-tion – the use of material resources beyond regional and global ecological capacities – must cease, because of the threat of imminent ecological collapse, does not allow for the self-serving gradualism of the weak forms of ecological modernisation discussed earlier.

Reflexivity and risk, anxiety and mistrust

Giddens has noted that the forms of reflexivity involved in the continual gener-ating of systematic self-knowledge do not stabilise the relation between expert knowledge and knowledge applied in lay actions. This is as true for scientific and technical systems as for sociology (to which Giddens was referring), for these systems also remain always at least one step away from the understanding which would control their impacts and are always on their way to creating new problems. Increased ecological awareness encourages recognition of the limits to our scientific comprehension of the physical world and therefore of the limits to our capacity to know and technically manipulate them. Our crude understand-ing of the interplay of biological systems and global climate is a good case in point.

However industrial modernisation has largely vanquished the traditional cultural forces which might control the abstract scientific appropriation of the environment or, more importantly, the impulse to transform nature (whether through biotechnology or *in vitro* fertilisation). At the same time, it has produced a new category of 'socio-technological' failures – such as Chernobyl and the ozone hole – which is unprecedented in its spatial and/or temporal reach, respecting no territorial borders and potentially affecting future generations. The global extension of the 'catastrophic capacity' of industrial modernisation is accompanied by the means to broadcast information about such disasters to populations which previously trusted expert systems and now become aware of these new risks, with their implications at the personal (cancer and death) and global (destruction of life on Earth) levels of existence.

As Giddens, Beck and others point out, the resultant disenchantment with science and technological change and the popular appreciation of the new risks they produce, has led to a transformation of public perceptions of 'progress'. Optimistic notions of progress, based in uncritical belief in the benefits of the scientific and industrial appropriation of nature, have now collapsed into anxiety and mistrust. Giddens argues that this new phase, which other theorists call 'postmodernity', is but an extension of modernity in process. 'We have not moved beyond modernity but are living precisely through a phase of its radicalisation', a period in which progress is 'emptied out by continuous change' (Giddens 1990: 51). However there is good reason to suggest that this underplays the discontinuities associated with cultural disenchantment with progress and particularly its handmaidens, science and technology. This disenchantment constitutes a radical departure from 'simple modernity' and signals the establishment of a new, more anxious phase of reflexive modernisation (Beck *et al.* 1994).

Those interpretations of ecological modernisation which are still embedded in notions of industrial progress,[19] albeit more cautious but still bearing an evolutionary sense of technological adaptation through reflexivity, do not address the extent of this corrosion of trust in 'simple' industrial modernity. They accept that modernisation has become *more* reflexive, but only in the narrow and instrumental sense of improving 'environmental efficiency', rather than in the broad and *reflective* manner of ecological critique which fundamentally questions the trajectories of industrial modernity. By contrast, strong ecological modernisation, therefore, also points to the potential for developing a range of alternative ecological modernities, distinguished by their diversity of local cultural and environmental conditions; although still linked through their common recognition of human and environmental rights and a critical or reflexive relationship to certain common technologies, institutional forms and communicative practices, which support the realisation of ecological rationality and values ahead of narrower instrumental forms.

In conclusion, the concept of ecological modernisation has been deployed in a range of ways – as a description of narrow, technological reforms; as a term for policy analysis; in reference to a new ideological constellation; and in reference to deeply embedded and ecologically self-conscious forms of cultural transformation – and bearing quite different values. As a result, there is a danger that the term may serve to legitimise the continuing instrumental domination and destruction of the environment and the promotion of less democratic forms of government, foregrounding modernity's industrial and technocratic discourses over its more recent, resistant and critical ecological components. Consequently, there is a need to identify the normative dimensions of these uses as either weak or strong, depending on whether or not such ecological modernisation is part of the problem or part of the solution for the ecological crisis.

Notes

1 I am grateful to John Dryzek, Robyn Eckersley, Boris Frankel and Paul James for comments on an earlier draft.
2 For instance, see Simonis (1988); Jänicke *et al.* (1992); Zimmermann *et al.* (1990); Weale (1992); Hajer (1995) and Andersen in this volume.
3 For example, see Vogel (1986,1990); Vogel and Kun (1987); Knoepfel and Weidner (1990); Vig and Kraft (1990); Yaeger (1991); Weale (1992); Feigenbaum *et al.* (1993); Wintle and Reeve (1994).
4 For instance, while transboundary problems such as acid rain and fallout from Chernobyl shaped environmental politics, policies and institutions in western Europe, they were of little consequence in Japan and irrelevant to 'frontier states' such as Australia, where preservationist conflicts over the impacts of primary resource extraction – agriculture, forestry and mining – on relatively-pristine environments predominated.
5 The NEPP has already undergone two four-yearly reviews, as required by legislation.
6 Consider the enormous gap between the technical capacities – which have been available for decades – to produce durable, safe, energy-efficient and largely-recyclable cars and the actuality to date.
7 Hajer's use of the term varies in its elasticity. As he extends his view of ecological modernisation to the point that it seems all-embracing in its cultural inclusivity, it becomes hard to see what bounds ecological modernisation as a discourse – a theoretical-methodological problem common to Foucauldian approaches to policy analysis. Perhaps it is therefore better to regard ecological modernisation instead as a metadiscourse or deep cultural tendency. It then becomes possible to read ecological modernisation back into the nineteenth-century movement for resource conservation and forward into the growing reflexivity of science and technology. That Hajer might want to add conservative and neoliberal opposition to state regulation to his list of the signs of ecological modernisation indicates some of the problems with his own ill-defined discursive approach to 'locating' ecological modernisation.
8 Towards the end of *The Politics of Environmental Discourse* (Hajer 1995), Hajer briefly touches upon an ideal form of ecological modernisation which he calls 'reflexive ecological modernisation'. This represents a cultural tendency rather than merely a policy discourse and stands in opposition to techno-corporatist ecological modernisation in its emphasis on democratic and discursive practices.
9 See Christoff (1996).
10 For instance, see Weale (1992: 78–9).
11 Both Weale and Hajer comment on the role which international forums, such as the OECD, have played in fostering ecological modernisation (EM) as a policy discourse. For instance, Weale claims: [T]he main bodies responsible for developing the ideology of EM were international organisations, who sought to use the new policy discourse as a way to secure acceptance of common, or at least

harmonised, environmental policies, the closest example being the EC. (Weale 1993: 209) He also discusses the evolution of new international environmental regimes, but does not integrate this discussion into his exploration of ecological modernisation (Weale 1992: chap. 7).

12 Similarly, Jahn (1993: 30, and in this volume) notes that data seem to indicate that neocorporatism has a positive impact on environmental performance and on antiproductionist politics. He comments that 'it seems reasonable to argue that the impact of neo-corporatist arrangements on both dependent variables is dependent upon the influence of new social movements and associated green and left-libertarian parties on established politics'. However, importantly, he also notes Offe's observation that the cost of corporatist arrangements is the marginalisation of non-organised interests, which is antithetical to the democratic principles of new social movements and of green politics.

13 Weale (1992: 31) and Hajer (1995: 280 ff.) suggest but do not explore such alternatives.

14 By 'ecological critique' I mean both the emergent scientific understanding of ecological needs which has evolved out of the biological and physical sciences, and the normative and non-instrumental (re)valuation of nature (including its spiritual and aesthetic aspects as these manifest in concern for preservation of species and ecosystems, 'wilderness' and visual landscape values). Both are elements increasingly dominant, motivating features of the environment movement in the late-twentieth century.

15 This section draws heavily upon Giddens' elegant long essay, *The Consequences of Modernity* (Giddens 1990).

16 Of course these 'newer' forms of trust in abstract systems may be related to 'premodern' forms of trust in cultural explanatory frameworks (religion, myth, etc.). They coexist with and interact in the process of identity formation with more direct forms which are essential in face-to-face communities and intimate social relations (as in families).

17 This is not to argue for a return to essentialised and romantic, exclusionary and parochial notions of 'place', such as have been central to the campaigns of certain environmental communitarians. While arguing for the need to recognise and preserve the specific place-bounded nature of ecological relations, it is also important to note the ways in which cultural notions of identity and 'place' have been irrevocably transformed by modernity as, globally, face-to-face communities are now infused by the informational attributes and other requirements of abstract exchange.

18 Its 'abstract' knowledge of nature remains based on research and investigation, on the international transmission of new scientific information among scientists, environmental managers and environmentalists, as well as (potentially) upon the recovery and reauthorisation of aspects of 'local' indigenous knowledge.

19 See Jänicke (1988); Hajer (1995: 33).

References

Amin, A. (1994) *Post-Fordism: a Reader*, Oxford: Blackwell.

Andersen, M. S. (1993) 'Ecological modernisation: between policy styles and policy instruments – the case of water pollution control', Paper delivered at 1993 ECPR Conference, Leiden.

Angerer, G. (1992) 'Innovative technologies for a sustainable development', in F. J. Dietz, U. E. Simonis and J. van der Straaten (eds), *Sustainability and Environmental Policy*, Berlin: Bohn Verlag.

230   *P. Christoff*

Beck, U. (1992) *Risk Society: Towards a New Modernity*, London, New York: Sage Publications.

Beck, U., Giddens, A. and Lash, S. (1994) *Reflexive Modernization: Politics, Tradition and Aesthetics in the Modern Social Order*, Cambridge: Polity Press.

Best, M. H. (1990) *The New Competition: Institutions of Industrial Restructuring*, Cambridge: Polity Press.

Carley, M. and Christie, I. (1992) *Managing Sustainable Development*, London: Earthscan.

Christoff, P. (1996) 'Ecological citizenship and ecologically guided democracy', in B. Dougherty and M. de Guis (eds), *Democracy and the Environment*, London, New York: Routledge.

Eckersley, R. (ed.) (1995) *Markets, the State and the Environment: Towards Integration*, Basingstoke, London: Macmillan.

Everett, M. (1992) 'Environmental movements and sustainable economic systems', in F. J. Dietz, U. E. Simonis and J. van der Straaten (eds), *Sustainability and Environmental Policy*, Berlin: Bohn Verlag.

Feigenbaum, H., Samuels, R. and Kent Weaver, R. (1993) 'Innovation, coordination and implementation in energy policy', in R. Kent Weaver and B. A. Rockman (eds), *Do Institutions Matter? Government Capabilities in the United States and Abroad*, Washington, DC: Brookings Institute.

Giddens, A. (1990) *The Consequences of Modernity*, London: Polity Press.

Gross, N. (1992) 'The green giant? It may be Japan', *Business Week*, 24 February.

Haeckel, E. (1870) *Natürliche Schöpfungsgeschichte*, Berlin: Reimer.

Hajer, M. A. (1995) *The Politics of Environmental Discourse: Ecological Modernization and the Policy Process*, Oxford: Clarendon Press.

Harvey, D. (1993) 'From space to place and back again: reflections on the condition of postmodernity', in J. Bird, B. Curtis, T. Putnam, G. Robertson and L. Tickner (eds), *Mapping the Futures: Local Cultures, Global Changes*, London, New York: Routledge.

Hatch, M. T. (1995) 'The politics of global warming in Germany', *Environmental Politics* 4, 3: 415–40.

Jahn, D. (1993) 'Environmentalism and the impact of green parties in advanced capitalist societies', paper delivered at 1993 ECPR Conference, Leiden.

Jänicke, M. (1988) 'Ökologische Modernisierung: Optionen und Restriktionen präventiver Umweltpolitik', in U. E. Simonis (ed.), *Präventative Umweltpolitik*, Frankfurt, New York: Campus Verlag.

—— (1990) *State Failure: the Impotence of Politics in Industrial Society*, London: Polity Press.

—— (1992) 'Erfolgsbedingungen von Umweltpolitik im internationalen Vergleich', *Zeitschrift für Umweltpolitik* 3, 2: 213–311.

Jänicke, M., Monch, H. and Binder, M. (eds) (1992) *Umweltentlastung durch industriell Strukturwandel? Eine explorative Studie über 32 Industrieländer (1970–1990)*, Berlin: Edition Sigma, Rainer Bohn Verlag.

Knoepfel, P. and Weidner, H. (1990) 'Implementing air quality programs in Europe: some results of a comparative study', *Policy Studies Journal* 11, 1: 103–15.

McKibben, B. (1990) *The Death of Nature*, Harmondsworth: Penguin.

Moore, C. A. (1992) 'Down Germany's road to sustainability', *International Wildlife* 16, September/October: 24–8.

OECD (1993) *OECD Environmental Data 1991*, Paris: Organisation for Economic Co-operation and Development.

Offe, C. (1992) 'Bindings, shackles and brakes: on self-limitation strategies', in A.

Honneth, C. Offe and A. Wellmer (eds), *Cultural-political Interventions in the Unfinished Project of Enlightenment*, Cambridge, MA: The MIT Press.

Simonis, U. E. (ed.) (1988) *Präventative Umweltpolitik*, Frankfurt, New York: Campus Verlag.

van der Straaten, J. (1992) 'The Dutch National Environmental Policy Plan: to choose or to lose', *Environmental Politics* 1, 1: 45–71.

Vig, N. J. and Kraft, M. E. (eds) (1990) *Environmental Policy in the 1990s: Towards a New Agenda*, Washington, DC: Congressional Quarterly Press.

Vogel, D. (1986) *National Styles of Regulation: Environmental Policy in Great Britain and the United States*, Ithaca, NY: Cornell University Press.

—— (1990) 'Environmental policy in Europe and Japan', in N. J. Vig and M. E. Kraft (eds), *Environmental Policy in the 1990s: Towards a New Agenda*, Washington, DC: Congressional Quarterly Press.

Vogel, D. and Kun, V. (1987) 'The comparative study of environmental policy', in D. Meinolf, H. N. Weiler and A. B. Antal (eds), *Comparative Policy Research: Learning from Experience*, New York: St Martin's Press.

Wallace, D. (1995) *Environmental Policy and Industrial Innovation*, London: Earthscan.

Weale, A. (1992) *The New Politics of Pollution*, Manchester, New York: Manchester University Press.

—— (1993) 'Ecological modernisation and the integration of European environmental policy', in J. D. Liefferink, P. D. Lowe and A. P. J. Mol (eds), *European Integration and Environmental Policy*, London: Belhaven.

Wintle, M. and Reeve, R. (eds) (1994) *Rhetoric and Reality in Environmental Policy: The Case of the Netherlands in Comparison with Britain*, Aldershot: Avebury; Brookfield, VT: Ashgate.

WRI (World Resources Institute) (1994) *World Resources 1994–95 – a Guide to the Global Environment*, New York, Oxford: Oxford University Press.

Yaeger, P. C. (1991) *The Limits of the Law*, Cambridge: Cambridge University Press.

Zimmermann, K., Hartje, V. J. and Ryll, A. (1990) *Ökologische Modernisierung der Produktion: Struktur und Trends*, Berlin: Edition Sigma.

10 Disciplining the market, calling in the state

The politics of economy–environment integration

Robyn Eckersley

Introduction

The question of how to integrate the economy and the environment is shaping up to be one of the major political battlegrounds at the start of the new century, as governments seek to give some kind of effect to the sustainable development rhetoric of the 1980s and 1990s. Although the environment movement and green political parties have been in the forefront of publicising and problematising issues of sustainability, they have no monopoly on the definitional stakes nor on the national or international policy debates. It is now a trite observation that on none of the sustainability questions of how, for whom and to what end is anything approximating broad social consensus to be found.

To recognise the essentially contested nature of the concept of sustainable development is necessarily to recognise that there is no universal or 'objective' measure by which to evaluate competing models of economy–environment integration. This applies not only to the overt political claims made by parties, movements and interest groups, but also to the policy analyses and prescriptions offered by 'professional' advisers to government – most notably economists – who are playing a key role in shaping both the research and political agendas relating to the selection of environmental policy instruments.

Moreover, shifting the discourse from 'sustainable development' to 'ecological modernisation' does not alter the fundamentally normative character of the environmental debate. If one of the hallmarks of ecological modernisation is the increasing reflexivity of policy-makers and the public in relation to the production and distribution of risks (see Christoff this volume), and the central concern is how to address the endemic problem of 'problem displacement', then we must still contend with the basic questions concerning how, for whom and to what end. In particular, how and to what extent should ecological costs be avoided and/or allocated in relation to different social classes, geographical regions, nations, non-human species and future generations?

This chapter seeks to look behind the different invocations of sustainable development and ecological modernisation, to uncover the competing normative and methodological assumptions and claims that are embedded in different problem definitions and associated policy prescriptions for economy–environment integration. Despite the novelty of the ecological challenge, most of these policy prescriptions continue to line up along the well worn political spectrum characterised by the degree of state 'intervention' in the economy. As William Leiss has explained,

> neither our market economy nor our methods of public decision making [...] were designed with environmental concerns in mind. Our political economy responds to those concerns on the basis of well established private interests, distribution of power, lines of authority, and ideologies. Its "instinctive" response to environmental concerns is to define problems according to the capacity of existing institutions to deal with them – in this case, by extending existing regulatory and price mechanisms to deal more adequately with pollution.
>
> (Leiss 1990: 96)

I have elsewhere argued that the green movement may rightly claim a considerable measure of originality and distinctiveness in relation to its *analysis* of environmental problems, and in relation to the particular *values* and *goals* it embodies (especially those associated with ecocentrism). However, with the possible exception of bioregionalism, the green movement has offered very little that is new in relation to *social organisation and institutional design*, at least at a general societal level as distinct from movement level (Eckersley 1992b: 31). In other words, most of the social, political and economic arrangements defended by green political theorists have already been mooted in the history of political thought, albeit with different problems and different political ends in mind. Indeed, even bioregionalism represents a special ecological adaptation of a confederal model of political organisation that has long been defended by anarchists.

To suggest that there is no uniquely 'green' institutional design for an ecologically sustainable society, is not to suggest that some institutional designs may not be more *conducive* to the development and maintenance of such a society than others. However, the matter is complicated by the fact that environmental protection is not the *only* goal of green political actors – and even less so of other political actors. This chapter seeks to explore the ways in which ecological norms and objectives are arranged in relation to *other* political norms and objectives, and how particular constellations of norms and objectives are expressed through particular institutional designs, decision rules and policy instruments. Once this general level of analysis is pursued, attention to more general political theories, albeit flavoured by ecological problems and concerns, becomes unavoidable.

Linking the debate about policy instruments with broader, competing political objectives also makes it possible to explain what appears to be a convergence of support among different political stakeholders (governments, industry, environmentalists) in relation to the use of particular environmental policy instruments. For example, the growing disillusionment with traditional environmental regulation and the increasing advocacy of market-based instruments (e.g. taxes, charges, tradeable emission permits) over the last decade should not be read as a sign of consensus among the major political stakeholders, if it can be shown that such stakeholders are defending the use of market-based instruments for different reasons. Indeed, I have shown elsewhere how market-based instruments have been pursued on both pragmatic and ideological grounds, to promote radical, moderate and conservative environmental causes (Eckersley 1995: 12).

It is possible to identify four major competing political orientations and associated institutional frameworks that have been staked out in the sustainable development debate. These are the privatised market economy (defended by free-market environmentalists); the macrodisciplined market economy (defended by green social democrats, for want of a better term); the socialised green economy (defended by ecosocialists); and the decentralised, self-reliant economy (defended by ecoanarchists, which includes bioregionalists). Each of these models rest on different analyses of the relationship between capitalism, the state and the ecological crisis, different social and ecophilosophical orientations, and different conceptions of democracy. The major points of contrast in these four models may be represented in the form of a simple matrix as depicted in Table 10.1.

Table 10.1 is intended to depict, in highly simplified fashion, the ways in which environmental problems are analysed and ordered *following their absorption into more general* political theories. It is important to emphasise that the four models represent political models rather than 'positive' economic models, although loose linkages with economic theories will be made. Through a critical analysis and comparison of the four general models, this chapter will suggest that the macrodisciplined market economy offered by green social democracy is likely to serve as the most conducive (but by no means trouble free) institutional framework for absorbing the ecological and social concerns of the green movement.

Ecoanarchism

Whereas the first three models depicted in Table 10.1 all neatly line up along the conventional political spectrum from right to left (determined according to the degree of state intervention in the market economy), the ecoanarchist model fits rather uncomfortably on this dimension. Indeed, the antistatist and decentralist response to the ecological crisis defended by ecoanarchists (Bookchin 1982; Sale 1985) is better accommodated along a state versus civil society/community control axis rather than a state–market axis. However, ecoanarchism is often defended as a viable model of sustainable development and so it is briefly included here for the sake of completeness.

Decentralist, community-based models of social organisation have enjoyed considerable popularity among many green supporters in both Europe, North America and Australasia – although not necessarily in their fully-fledged anarchist dress. Indeed, many greens (and certainly most environmental campaigners) acknowledge that local community initiatives, while important, are unlikely to stand alone as a total response to the ecological crisis – especially when set

Table 10.1 Four models of political orientations and associated institutional frameworks

	Cause of ecological problems	Role of state	Economy	Form of democracy	Wealth distribution
Free market environmentalism	Incomplete allocation of private property rights	Privatise environmental assets and wastes; leave environmental decision making to private contracting parties	Private capitalist; deregulated	Representative	Reward according to effort; wide income disparities; minimal welfare services
Green social democracy	Primarily 'market failure' but also 'state failure'	Market to provide primary resource allocation mechanism but state to provide just income distribution and sustainability planning	Private capitalist; heavily regulated; some state owned enterprises	Representative; participatory planning	Variation in income levels but with extensive welfare services and guaranteed adequate income support
Ecosocialism	Capitalism	State to provide economic, social and ecological planning	State owned enterprises; small private sector	Representative; participatory democratic planning	Relatively egalitarian (i.e. some variation within a relatively narrow band)
Ecoanarchism	Social hierarchy; lack of 'human-scaled' institutions; lack of local self-determination	Nation state rejected or by-passed; communities to organise on a local basis to meet their own needs from local resources	Self-reliant, local co-operative enterprises; unclear as to role, scale and operation of market	Direct, participatory	Relatively egalitarian in intent but unclear as to mechanisms

against the backdrop of the communications and transport revolutions, increasingly mobile labour and capital, the growing transboundary nature of many ecological problems and a global system made up of multinational corporations and sovereign states.

Indeed, many decentralist, community-based prescriptions are notably unclear as to the mechanisms for redistributing wealth; compensating for natural resource endowment disparities; or dealing with transboundary environmental problems between local communities, within regions, or between regions. After all, ceding political and economic autonomy to local communities within confederated bioregions provides no guarantee that such communities will choose development paths that are ecologically sustainable. Indeed, local communities can often be parochial and susceptible to domination by local vested interests. Moreover, many ecoanarchist models tend to rely on voluntarist ecological solutions by 'right-minded' local groups and citizens on the basis of idealised participatory and consensual decision-making structures that gloss over the intricate dynamics of power, expertise and personality, and require degrees of patience and commitment that none but a small minority of dedicated activists seem willing to provide.

Ecoanarchist theories do, however, provide a valuable critical perspective on, and supplement to, statist models and, in some cases, can provide important prescriptions for dealing with particular ecological problems (such as the management of watersheds). I have critically examined these theories elsewhere (Eckersley 1992a,b: chap. 7) and do not propose to rehearse this discussion in any further detail here.

Free-market environmentalism

Free-market environmentalism represents the most concerted defence of the virtues of private property and the market in solving environmental problems (Anderson and Leal 1991; Moran *et al.* 1991). Proponents of free-market environmentalism argue that environmental problems arise not from the rational pursuit of profit for private gain in respect of free environmental resources, but rather from the lack of a clear specification of exclusive, transferable and enforceable property rights in respect of such resources. 'The problem', as Ackroyd and Hide (1991: 189) succinctly explain, 'is the lack of the very institution that lies at the heart of the free enterprise system', namely, private property. The reason we have polluted beaches, oceans and air is explained by the fact that there are no private-property rights in, and hence no markets for, unpolluted beaches, oceans and air. If the environment were fully owned by private stakeholders, then there would be no need to impose environmental regulations, because the bargaining and enforcement of the relevant private-property rights through the common law would ensure that either the appropriate costs were internalised or the appropriate compensation paid by the private parties involved (Helm and Pearce 1991: 8). In short, the solution to the 'tragedy of the commons' is the privatisation of the commons.

Free-market environmentalism, which is often associated with the work of Coase (1960) and the Chicago school of economists, is defended as a more robust *alternative* to government regulation and taxation. Indeed, free-market environmentalists seek, wherever possible, the *abdication* of state control over the operation of markets (except in relation to the enforcement of private-property rights). It is this latter aspect that distinguishes free market environmentalists from neoclassical environmental economists, who recommend government 'correction' of 'market failure' by means of Pigovian taxes – after Pigou (1920) – and other market-based instruments, including tradeable permits. The latter school are market-*oriented*, but they do not advocate *free* markets (Jacobs 1992: 2). The distinguishing feature of free-market environmentalism, as Jacobs correctly points out, is not the creation of property rights per se, but rather the removal of state control over the level or amount of environmental exploitation carried out by private-property owners (Jacobs 1993).

Free-market environmentalists attribute most environmental problems not to market failure but to 'state failure'. Politicians, bureaucrats and other public officials are assumed to act in self-interested ways rather than pursue general social welfare. Centralised 'political choice' is therefore condemned as grossly ill-informed, inefficient and distorting (through self-interested power trading, 'pork barrelling' and so forth). Indeed, proponents of free-market environmentalism even go so far as to maintain that their approach is *more democratic* than 'state environmentalism' on the grounds that it decentralises and depoliticises environmental resource-allocation decisions; enables the creation of a personal stake in the environmental asset that is the subject of the property right; and provides a superior mechanism for addressing the informational requirements of environmental decision-making. In effect, in defending decentralised private choice over centralised 'political choice', free-market environmentalists are seeking to replace the cumbersome notion of 'citizens democracy' (read centralised, 'command-and-control' approach to environmental management) with the efficient and decentralised 'business democracy' of the market place in the allocation of environmental resources.

On one view, the case for free-market environmentalism has a certain elegant simplicity and consistency. However, this is achieved by assuming the existence of competitive markets and shielding from view a range of contentious political questions, the resolution of which are essential to the political acceptability and environmental effectiveness of such a full-blooded property-rights approach. These questions relate to the democratic, distributional, informational and environmental 'failures' of private-property regimes.

A central problem with free-market environmentalism is its refusal to countenance the distinction between entrepreneurship and citizenship. This refusal may be partly traced to the 'rational-actor model' upon which free-market environmentalism rests. According to this model, all individuals are assumed to be well informed in relation to their own interests and concerned only in making decisions which maximise their own particular set of preferences. The issue of the provision of public goods is presented as a debate about informed individual choices (market) versus external, imperfectly informed coercion (state). The upshot is an impoverished interpretation of democracy as 'the freedom to make one's own economic decisions, uncoerced by the state'. Such a narrow interpretation of democracy leaves no room for the possibility of the *transformation* of individual preferences through reasoned political dialogue about generalisable (as distinct from merely individual) social and environmental interests. The market is only one of a range of social institutions which may be made available to affected parties to negotiate different courses of action in relation to the environment.

Free-market environmentalists are also conveniently silent on the crucial question of *how* environmental property rights are to be acquired and distributed. Many of the arguments simply assume that trading regimes, organisations and entrepreneurs will somehow spontaneously emerge and that governments will merely enforce contracts when conflicts arise. Yet the methods and criteria of the acquisition of environmental property rights are crucial questions that determine *who* is able to participate in the new markets, the rules of participation (i.e. whether bargaining is free and fair), and what the environmental outcomes might be. Free-market environmentalists maintain that well enforced environmental property rights make co-operative exchanges possible between economic actors. However, as Schmidtz (1993: 531–2) points out, it is the reciprocity that is the ability of parties to conditionalise offers, not the property rights per se, which actually effects the exclusion and the allocation of resources. Clearly, property rights are a necessary but not sufficient condition for reciprocity. Social norms of trust and honesty, relative parity of bargaining power and non-prohibitive transaction costs are also required before voluntary, reciprocal exchanges may take place.

Now, it may be possible to devise initial conditions for private bargaining among the holders of environmental property rights in ways that will ensure environmentally informed and fair bargaining between the parties. However, such a regime could not be established without significant resource transfers or adjustments effected by the state, a move that many free-market environmentalists would deem an unwarranted infringement of the liberty of existing holders of property rights. Yet, in the absence of such adjustments, the property-rights approach is one that will ensure that not everyone is 'free to participate' or 'free to choose' in these new markets, just as not everyone is 'free' to participate on an equal footing in any market. Non-property holders will be totally excluded from the environmental decision-making process, even though they may be affected by private bargaining between property holders. The putative 'depoliticisation' of resource-use conflict defended by free-market environmentalists must therefore be seen, in large measure, as a thinly-disguised endorsement of the existing distribution of economic and political power between, on the one hand, resource-extractive industries and other economic actors who are keen to exploit the environment and, on the other hand, environmental groups who are keen to protect the environment. Indeed, the long term consequence of zealously pursuing the 'privatisation of the environment' is likely to be the intensification of the already wide gap between the propertied and the propertyless and the transmutation of the ideal of representative democracy to that of 'commercial democracy'.

Moreover, the ability of individuals to participate in reciprocal and ecologically-informed private negotiations is a function not only of the purchasing power and information at their disposal but also of their individual preferences, which may or may not accord with ecological concerns. Many environmental problems are complex, uncertain and not always apparent to the untrained eye. In the absence of specialist information as to the wider consequences (in space and time) of particular environmental decisions, even the greenest of property holders may unwittingly expose others to considerable risks of harm. Moreover, we have seen that those concerned to protect environmental values do not ordinarily have the same bidding power as those who are eager to pursue the economic exploitation of the environment. And even if we were to assume that private parties have 'perfect environmental information' and bargaining parity, private negotiations and the enforcement of private-property rights will not always result in improved environmental quality, because parties may prefer to accept monetary compensation rather than seek the cessation of the environmentally degrading activity.

Finally, only a small proportion of the wide-ranging values associated with environmental 'goods' can be 'captured', quantified and valued in economic terms. This is especially so in relation to the values of general ecosystem services and biodiversity. Even in those cases where particular species of wildlife do have commercial value (for their meat, hides, 'scenery value', or as 'game' for hunters), the creation of property rights in such species effectively confers on the property holders the freedom to trade off ecological values and species against other uses. This is fully consistent with the free-market environmentalist's casting of environmental concerns as mere matters of private tastes. The environment will always be vulnerable to being outbid by competing uses.

Free-market environmentalism rests on an ethical and methodological individualism and proceeds on the basis of a deep distrust of any form of 'political choice'. If the search for the integration of ecological and environmental decisions is reduced to optimising allocative efficiency, then free-market environmentalism has something to offer, at least in relation to *competitive* markets. However, if integrating the environment and economy is about minimising problem displacement (across social classes, species, generations and geographic regions), then the prescriptions of free-market environmentalism must be treated with great scepticism by greens. Indeed, any approach that defends the market as the best mechanism for 'deciding' the allocation of environmental goods (and bads), carries with it an inbuilt bias against the poor, against future generations and against non-humans. The market may well provide the best means for enabling an *efficient allocation* of resources, but it is systematically blind to questions of participation, just distribution and ecological sustainability (Daly 1991). It is precisely because so many environmental problems transcend all manner of boundaries – human and non-human, spatial and temporal – that more collective forms of social choice are required. On all counts: ecological sustainability, intra- and intergenerational equity and democratic participation; free-market environmentalism is worse than the status quo.

Green social democracy

Green social democracy is admittedly somewhat of an amorphous term that could doubtless withstand further political subdivision. Historically, the label 'social democrat' was used to distinguish the peaceful, reformist and democratic socialist from the violent and revolutionary socialist. Nowadays, it is generally taken to refer to a left-of-centre political orientation, usually allied with the labour movement, that seeks to prevent, cushion, or ameliorate the dislocations and social hardships generated by capitalism, in order to create a more socially-egalitarian society. However, the introduction of the 'green' qualifier radically transforms these broad objectives.

Like other social democrats, green social democrats look to the state to play a key role in preventing and/or reversing the privatisation of benefits and socialisation of costs that characterise the market economy. However, unlike other social democrats, *green* social democrats also seek to recontextualise this traditional social-democratic quest for greater social equity by ensuring that it is no longer pursued at the expense of the environment. In other words, distributional struggles are to be bounded by, or take place within the context of, ecological constraints.

Unlike free-market environmentalists, green social democrats see the state as playing a vital role in correcting 'market failure', and acting as environmental protector and guardian of generalisable and long-term interests. Green social democrats accept the institutions of private property and the market as the basic mechanisms of wealth creation, but insist that the state ought to play a much more active role in disciplining and channelling market transactions in ways that produce environmentally and socially-beneficial outcomes – see, for example, Daly and Cobb (1989); Zarsky (1990); Jacobs (1991); Ekins (1992). Their political and economic program seeks to use the state as the vehicle for the democratic, collective determination of sustainability parameters within which the market *and* the state may operate. In this way, they hope to deliver macrostability, sustainability and equity while preserving maximum microfreedom and local diversity. As Daly and Cobb (1989: 48–9) explain:

> If one favors independence, participation, decentralised decision making, and small- or human-scale enterprises, then one has to accept the category of profit as a legitimate and necessary source of income. There is plenty of room to complain about monopoly profits, but that is a complaint against monopoly, not against profits per se [...] If one dislikes centralised bureaucratic decision making then one must accept the market and the profit motive, if not as a positive good then as the lesser of two evils [...] We

have no hesitation in opting for the market as the basic institution of resource allocation.

However, Daly and Cobb also acknowledge that the price mechanism merely claims to ensure an optimal *allocation* of scarce resources (an efficiency issue – and even this is considerably undermined by the lack of perfect competition in the real world), not an optimal *scale* of resource use (an ecological–ethical issue), nor an optimal *distribution* of resources (a social–ethical issue) (Daly and Cobb 1989: 145–6).

The three most significant ecological shortcomings of the market that warrant state intervention are the familiar and interrelated problems of firms 'externalising' ecological costs, the indiscriminate growth dynamic of capitalism and the problem of discounting the future. These three problems are all interrelated by-products of 'rational' (i.e. profit maximising) market behaviour.

The first of these problem is well known and is usually attributed to the fact that most environmental goods (e.g. clean air or the waste assimilation services of ecosystems) have a zero price and, in the absence of government regulation, there is no incentive for firms to economise on their use and no disincentive for firms to pass these costs on to third parties and/or the general public. Green social democrats are concerned to alter the incentive structure facing economic actors to ensure that these externalities are avoided or drastically reduced. As will be explained below, environmental taxes are not necessarily the most appropriate means of *preventing* externalities.

The second problem (which is partly related to the first) arises because the market economy decrees that firms 'grow or die'. This imperative for continual economic growth does not respect physical limits to growth or the notion of ecological carrying capacity. As Pearce and Turner (1990: 24) point out

[m]odern economics lacks what we call an *existence theorem*: a guarantee that any economic optimum is associated with a stable ecological equilibrium [...] The Pareto optimality of allocation [...] is independent of whether or not the scale of physical throughput is ecologically sustainable.

The appropriate role of the state, then, is to introduce ceilings, bottom lines, minimal standards and general sustainability limits to *contain and channel* market transactions in ways that limit the rate or material-energy throughput, so that the economy operates safely within the carrying capacity of ecosystems. Jacobs (1991: 120) refers to this as 'sustainability planning', which requires the development of primary indicators measuring environmental capacity (e.g. soil, water and air quality; atmospheric carbon dioxide concentrations; species diversity) and secondary indicators measuring the economic activities that cause changes in the primary indicators (e.g. emissions, chemical use, resource exploitation). On the basis of such information, the state is then in a position to determine what are likely to be the most effective instruments to curtail environmentally degrading activities to levels appropriate to protect ecosystems and biodiversity.

The third problem of 'discounting the future' means that market rationality continually gives priority to short-term interests over long-term interests – a practice that creates a structural bias *against* future generations (of both humans and non-humans). Indeed, high interest rates can serve as a strong disincentive to sustainable resource use, by encouraging the liquidation or depletion of both renewable and non-renewable resources and the movement of the capital thereby gained into more profitable ventures. Again, the appropriate response of government here is to enforce limits, quotas and other regulations that ensure sustainable harvesting practices in respect of natural resources, fisheries and agriculture. The problem of discounting can be also be tackled more generally by managing the economy in ways that keep inflation and interest rates low.

In terms of its compatibility with economic theory, green social democracy clearly owes more to the emergence school of ecological economics (of which Daly is a major proponent), than to the Coasian or Pigovian schools within neoclassical environmental economics. However, in fleshing out the ways in which the state might manage the market to defend environmental values and integrate economic and environmental decisions, green social democrats support the more widespread use of market-based instruments (e.g. green taxes, charges, deposit-refund schemes, subsidies and tradeable emission permits) in circumstances where the state retains the power to control sustainability parameters. So, for example, green social democrats generally support the judicious use of tradeable property rights in certain appropriate cases (such as point-source pollution in a competitive market setting) when combined with sustainability limits set by the government (such as strict emission ceilings in airshed and waterways) and fair terms of allocation and trade.

The utilitarian model of democracy on which environmental economics rests is also too 'thin' for most green social democrats. Environmental economists reduce environmental values to the environmental preferences of individuals, which are measured by constructing hypothetical markets in respect of certain environmental goods. These individual environmental preferences are then monetised, aggregated and incorporated into cost-benefit analysis to provide environmental policy-makers with information concerning the welfare gains to be made from different courses of action. The democratic defence of this method is that monetisation not only registers individual environmental preferences, it also reflects the *depth* of feeling contained in each preference expression (Pearce *et al.* 1989: 55). However, this method does not enable communication and debate among individuals. Nor does it follow that the sum of individual preferences will necessarily lead to outcomes that will advance the social or ecological community conceived *as a whole* or conceived as *different communities of interest.* These are interests that are most appropriately debated and determined in the public sphere *as generalisable interests* – ideally by a 'rationalised speech community' of the kind defended by Habermas (1984).

In any event, the objective of environmental economics is to ensure that externalities are internalised to arrive at an 'optimal' allocation of resources. By raising the price of environmental goods (whether directly or indirectly), environmental economists assume that we can overcome the discrepancy between the internal or private costs of production and the external or wider social costs of production. That is, the final (higher) prices are assumed to reflect the 'true' costs of production. Although the techniques for measuring these external costs have undergone considerable development in recent years, it is still extremely difficult to determine the wider social costs of production with any degree of accuracy in order that these costs be properly reflected in prices (this is especially so in the case of greenhouse-gas emissions). In any event, the purpose of government intervention from the point of view of green social democracy is not to ensure that environment costs are incorporated into prices to 'optimise' the efficiency of environmental resource allocation, but rather that environmental costs are not generated in the first place.

Here, the effectiveness of environmental taxes in reducing environmental damage will depend, among other things, on the consumer responsiveness to price changes in relation to particular goods and services. If demand is highly inelastic, higher prices will not lead to a significant drop in output (although it may generate welcome revenue). As 'target tools' for achieving restrictions in, say, pollution or greenhouse-gas emissions, environmental taxes (such as a carbon tax) can be very blunt instruments, because demand for energy is relatively inelastic. Environmental taxes can also be very indiscriminate instruments from the point of view of firms and consumers: they do not account for the different abatement costs facing different users and they are socially regressive (i.e. the higher prices resulting from the imposition of such taxes impact disproportionately on the poor; moreover, regressivity increases with inelasticity of demand).

Environmental taxes are, of course, only one among a range of different economic incentives that are available to governments to protect the environment. Moreover, green social democrats are not ideologically wedded to any particular economic instrument or policy measure. Rather, the concern is to use the most appropriate measure to achieve the objectives of sustainability, efficiency and social equity while ensuring public participation in the negotiation of targets and timetables. The point, in other words, is not the narrow Pareto optimality sought by environmental economists, but the explicit political negotiation of objectives in the public sphere, in the face of full information of the range of costs and benefits of different environmental policy options over different time periods.

Moreover, in determining the question of appropriate scale and carrying-capacity allowance, present and future generations of both the human and non-human world have to be accounted for. (Unlike most environmental economists (e.g. Pearce *et al.* 1989), green social democrats are less prepared to countenance the idea of substitutability of natural for human-made capital in discussions of intergenerational equity.) This can be given effect through much more comprehensive environmental-impact-assessment procedures that move beyond the current ad hoc, case-by-case appraisal of only large scale, environmentally 'significant' projects. These more comprehensive project-assessment procedures would be designed to operate within an overall strategy that is concerned to reduce the cumulative environmental impact of industry, agriculture, transport and urban development.

The green social democratic model extends the traditional social-democratic model of the welfare state by developing an ecologically-revised understanding of general welfare. The traditional welfare state has been dependent on a rising stock of wealth in order to carry out its redistributive functions; and it has therefore been keen to encourage *any kind* of growth and, consequently, it has been reluctant to introduce any measures that might be seen to compromise private capital accumulation (e.g. strict environmental standards).

In contrast, green social democrats would argue that this is a very narrow and short-sighted definition of general welfare, which ignores the inequitable distribution of environmental costs and benefits (in both space and time) generated by indiscriminate growth. By relying on more appropriate macroeconomic, welfare and environmental indicators (rather than indiscriminate gross domestic product figures), green social democrats seek to develop fiscal and monetary policies that encourage only 'green' or 'qualitative' growth for redistribution. This is, of course, a much more challenging task and in the short term there is every reason to believe that significant compromises will have to be struck between the demands of wealth generation, wealth redistribution and sustainability. There are also many challenges concerning administration and public participation that remain to be addressed. I will return to these vexed questions again in the conclusion, in the context of a discussion of the ecological challenge to bureaucratic administration.

Ecosocialism

Like green social democrats, ecosocialists also seek to give expression to the goals of ecological sustainability, social equity and participatory democracy, but they differ over how these values are best institutionalised. In particular, ecosocialists envisage a much greater role for the state in both economic and environmental planning and management as compared to green social democrats. Whereas green social democrats defend an increasingly regulated and ecologically disciplined (but still recognisably capitalist) market economy, ecosocialists argue that a more significant movement away from capitalism is required in order to overcome the social and ecological contradictions generated by market rationality – see, for example, Gorz (1980); O'Connor (1989); Ryle (1988); Weston (1986); Williams (1983); Pepper (1993). In place of the piecemeal, regulatory and essentially 'market-correction' model defended by green social democrats, ecosocialists argue that we must go beyond merely 'managing capitalism' by introducing more comprehensive state economic planning and greater public ownership of capital assets, while managing a much smaller and more heavily circumscribed private sector (Ryle 1988: 48).

Ecosocialists argue that for as long as the capitalist market economy is retained, the state will remain dependent on economic growth and a profitable private sector to fund its social and ecological reforms. They argue that it is not enough to institute new environmental indicators, a new national accounting system and new mechanisms of wealth distribution but otherwise leave the capitalist market system intact. It is also necessary to alter the institutions of capitalism, particularly the organisation and control of finance and investment. In short, ecosocialists stress the need to reorganise *production*, not just wealth distribution. Although there is considerable variation in terms of detail among the different economic programs outlined by ecosocialists, they generally envisage a combination of state and local democratic planning; democratically-controlled public enterprises; state regulation of the financial sector; self-managing worker co-operatives; and an informal or 'convivial' sector. Most ecosocialist programs also envisage the continuation of a small business sector, although profit accumulation and size are to be strictly controlled.

The claimed superiority of ecosocialism over green social democracy lies in that the state would no longer be fiscally parasitic on private capital accumulation to fund its social and ecological reforms. The upshot is that the contradictions of capitalism would, for the most part, be eliminated rather than simply 'managed' by the state. A democratically-planned economy is also defended as being better placed to (i) provide goods and services on the basis of need rather than purchasing power; (ii) avoid or minimise the negative externalities generated by market behaviour; (iii) iron out excessive social and regional inequalities; (iv) ensure that the scale of the macroeconomy respects the carrying capacity of ecosystems (unlike a market economy, a planned economy has no inbuilt imperative to grow); and (v) generally take a broader and longer-term view of the collective needs of present and future generations of both humans and non-humans (i.e. unhampered by the need to appease the immediate interests of private capital).

Ecosocialists argue that the ecological and social problems that have beset planned economies, can be attributed to a range of interrelated factors, which are neither necessary nor desirable aspects of a democratically-planned economy. These include rigid centralised control, single party bureaucratic rule, the absence of a free flow of information, the absence of an informed citizenry and popular participation, a commitment to industrialisation and high growth rates and a determination to 'catch up' with the West in terms of technical development and military might, as part of a perceived need to bolster 'national security' (O'Connor 1989). O'Connor (1989: 96) explains, 'in all socialist countries the major means of production are nationalised although not yet socialised, i.e. there is no strong tradition of democratic control of the means of life'. According to this argument, twentieth-century communism must be seen as an aberration rather than as an example of the inherent tendencies or 'logic' of a planned economy. If we remove all the repugnant features of these economies (as identified above) and ensure that production is properly 'socialised' rather than simply 'nationalised', then, so the argument runs, democratic self-management can emerge as a reasonably feasible option. As previously noted, a planned economy does not need to grow in the way that a market economy does. Moreover, we have seen that ecosocialists have explicitly rejected the path of indiscriminate economic growth and have embraced political pluralism, public accountability, freedom of information and widespread public participation in economic planning.

However, just as the theoretical defence of the market economy is based on the extremely restrictive set of assumptions of perfect competition, the theoretical defence of a democratically-planned economy is based on the naive assumptions of, *inter alia*, full information and complete trust between principal and agent. As Elster and Moene (1989: 4) observe, 'central planning would be a perfect system, superior to any market economy, if these two resources [full information and complete trust] were available in unlimited quantities'. The lack of continued availability of these two important 'resources' accounts for much of the inter-agency competition, displacement of responsibility and corruption that has characterised the state agencies of existing centrally-planned economies. Of course, these problems are not unique to bureaucracies in centrally-planned economies (i.e. they also apply to bureaucracies in liberal parliamentary democracies); however, such problems are likely to be magnified considerably by virtue of the expanded role played by state agencies in the ecosocialist economy.

The theoretical defence of a predominantly planned economy also assumes that it is possible to coordinate successfully a range of different state agencies in accordance with a 'common plan'. This presupposes (i) that it is possible to reach a social consensus on a common plan, and (ii) that each agency charged with implementing aspects of the plan will interpret it in a uniform way. As Dryzek points out, a teleological or goal-directed social system requires not just a consensus on values, but a *continuing* consensus on values if it is to produce, consistently and effectively, the desired common good. This arises from the fact that 'the administered structure cannot waver in its commitment, for it is only that commitment which can keep the system on course' (Dryzek 1987: 106); see also McLaughlin (1990: 92–3). (Again, this same criticism can also be applied, with somewhat less force, to the green social democrat's case for sustainability planning, which assumes a continuing democratic consensus.)

For all their caricatures of so-called command-and-control regimes, public-choice theorists may reasonably call on their detractors to demonstrate that the large-scale planning and co-ordination of economic activity will not lead to an authoritarian, hierarchical, planning and control structure. The celebrated convenience of the price mechanism, despite its many problems, is that it provides a relatively decentralised and seemingly depoliticised method of information processing and resource allocation that is responsive to consumer preference (at least in competitive markets). Moreover, its very 'invisibility' (that is, the absence of a deliberative steering system) has helped to maintain its political legitimacy to the extent that it obviates the need for ongoing social debate and consensus as to the merits or demerits of resource-allocation decisions. In contrast, state economic planning is more visible, discretionary and therefore more contestable than the impersonal, self-adjusting price signals of the market. In short, collective planning attracts criticism and debate as to the desirability of alternative courses of action. This is not necessarily a bad thing – indeed, it is precisely the kind of democratic debate that ecosocialists wish to substitute for the

impersonal signals of the market. Nonetheless, it raises a very familiar and difficult tension for those who wish to defend the path of ecosocialist pluralism: the
more the state replaces the market with a series of economic plans, the more it
needs to facilitate wide-ranging community consultation and consensus to maintain legitimacy. Yet the more ambitious the plan, the more reluctant the central
co-ordinating agency will be to tolerate and incorporate criticism that might
dispute its appropriateness or block its smooth implementation.

Nor is it clear whether the state apparatus has the capacity and flexibility to
redress the complex, systemic and increasingly transboundary nature of many
ecological problems – problems that throw down immense challenges to conventional administration and bureaucratic rationality. Both ecosocialism and green
social democracy place additional burdens of information gathering, assessment
and planning on the state; yet they are vague as to *how* these responsibilities might
be effectively discharged in an open, participatory and consensual manner.

In particular, the establishment of sustainability constraints will be an extremely challenging process in view of the high degree of scientific uncertainty
surrounding ecological problems, which are, as Dryzek succinctly points out,
inherently complex, non-reducible, variable, uncertain, spontaneous and collective in nature
(Dryzek 1987: 26–33). There is also a growing literature pointing to a fundamental incompatibility between traditional bureaucratic rationality and ecological
rationality (defined by Dryzek, somewhat narrowly, as the ability to manage
ecosystems to ensure that they provide consistently and effectively the good of
human life support). Indeed, some observers have suggested that the environmental crisis has given rise to an institutional crisis (Bartlett 1990: 82). Bureaucratic rationality is concerned to devise efficient means for achieving stated ends,
through problem decomposition and allocation, rule bound behaviour, speciali-
sation and routinisation and a top-down chain of command (Dryzek 1987: 88–
109). The rigid nature of traditional bureaucratic structures does not allow for
sufficient interactions *across* the boundaries of administrative subsets, which often
leads to problem displacement and an inability to respond swiftly and creatively
to changed circumstance (especially negative feedback).

It seems that the hard-nosed realism of the penetrating ecosocialist analysis and
critique of capitalism is not applied with the same vigour in the ecosocialist
assessment of the capacity of the state bureaucratic apparatus to carry out the
far-reaching responsibilities of planning and managing sustainable development.

The rocky road ahead

Most of the discussion so far has been concerned to explain and compare the
claims of each model and to defend the green social democratic model – not as
the *ideal* model, but as the *least-problematic* one. However, relatively little attention
has been given to the politics of implementation, to the sources and forms of
political mobilisation of, and opposition to, the green social democratic agenda.
After all, the implementation of this agenda is dependent on the formation of a
political consensus or effective majority that accepts the desirability and urgency

of sustainability planning. It also requires a political understanding and consensus on sustainability thresholds or ecological limits and on the measures to enforce such thresholds and limits.

In examining the politics of implementation, the problem for green social democrats is not simply the lack of an effective political majority to mandate sustainability planning, attention must also be directed to the complex and highly-uneven way in which competing political demands are mobilised, communicated, turned into policy commitments, and given legislative and administrative form in liberal democracies. This includes the role of scientists in 'detecting' and explaining ecological problems; the role of the media in framing problems and frequently simplifying and overdrawing conflict in complex environmental debates; and the general resource and power disparities of the major protagonists, particularly the role played by powerful oppositional forces (e.g. political parties, organised industry).

Although the green social democratic agenda is likely to have a greater general appeal than the ecosocialist agenda, insofar as it seeks to work more *with* the grain of the market economy, this in itself provides no assurance of the popularity or success of green social democracy. The green social democratic agenda seeks to impose a new range of constraints on private economic activity that will affect the short-term profitability of some industries and the long-term viability of others. Resistance from private capital and organised labour is therefore likely to be strong. Given the increasing international mobility of capital, the threat of a capital flight or strike, and to a lesser extent labour unrest, will significantly constrain the extent and pace of implementation of the green agenda. Here we return to the brute fact – which is central to the ecosocialist's case – that a green state will remain fiscally parasitic on private capital accumulation to fund its reforms. Far from disciplining the market, the green state – like all states in capitalist societies – is likely to be disciplined *by* the market.

A related set of challenges concerns the green social democratic objective of zealously pursuing a policy of redistributive justice, while simultaneously encouraging a contraction in ecologically-degrading economic activity. How, it might be asked, could a green state afford to fulfil its universalist promise of adequate income support for everyone, when it is also forcing the dismantling or contraction of a wide range of environmentally-degrading industries? Here, the success of green social policy will largely turn on the extent to which a green state will be able to facilitate a shift towards qualitative or 'post-material' growth, in accordance with a new green index of economic welfare – a feat that will require far-reaching changes in the present pattern of economic activity.

We have also seen that effective sustainability planning will require a considerable shake-up in the ordering and management of state agencies. The successful implementation of sustainability planning will require a considerable reordering in the traditional ministerial and departmental pecking order, so that environment ministerial portfolios are no longer regarded as junior portfolios and state departments of the environment are no longer seen as less important than central economic, trade and industry departments. Sustainability planning must apply to both private and public enterprises, as well as to state agencies, so that ecological considerations are thoroughly integrated into executive, administrative and judicial decision-making. However, we have also seen that the task of integrating environmental and economic decision-making requires a new (and yet to be developed) administrative structure that is considerably more flexible, open, collegial and reflexive than the present one (which still possesses many of the trappings of Weber's traditional-bureaucratic model).

Green social democrats must also contend with the critique of the welfare state, in particular the problems of dependency and alienation created by well-intended but ill-advised and poorly administered schemes designed to provide income support. It is somewhat of an irony that the greens are seeking to heap new responsibilities on the state at a time when there is growing scepticism about the ability of the modern welfare state to satisfactorily discharge its many contradictory functions. Resorting to the rhetoric of the 'enabling state' (whether for the poor or for the environment) is unhelpful in the absence of new participatory administrative models.

Finally, the green social-democratic agenda requires not simply that society's subsystems be capable of 'reading' ecological problems and responding to them in ways that minimise problem displacement; it also requires broader cultural changes and a deeper general understanding of ecological relationships so that, for example, practices such as waste minimisation, material-energy efficiency, recycling and use of public transport are considered to be 'civic virtues'.

Of course, all of the above requirements represent a very tall order, especially in our 'decentred', fragmented, mass society where there is an irreducible plurality of values and a culture of consumption and privatised social relations that leaves little time or incentive for political engagement or the cultivation of civic virtues. Whether these circumstances will continue to hinder the articulation of generalisable interests is an open question. Nonetheless, green social democrats are right to argue that the state remains the most appropriate, and certainly the most powerful, institution 'through which society can exercise some leverage upon itself' (Lacey 1991: 2) – especially in terms of reorganising and regearing its steering mechanisms along ecologically sustainable lines. Ecosocialists may well have the correct *analysis*; however, in the foreseeable future at least, green social democrats are more likely to build the necessary political consensus to develop a socially just and ecologically sustainable society.

References

Ackroyd, P. and Hide, R. (1991) 'A case study – establishing property rights to Chatham Islands' Abalone (Paua)', in A. Moran, A. Chisholm and M. Porter (eds), *Markets, Resources and the Environment*, North Sydney: Allen & Unwin.

Anderson, T. L. and Leal, D. R. (1991) *Free Market Environmentalism*, San Francisco, CA: Pacific Research Institute for Public Policy.

Bartlett, R. V. (1990) 'Ecological reason in administration: environmental impact assessment and administrative theory', in R. Paehlke and D. Torgerson (eds),

Managing Leviathan: Environmental Politics and the Administrative State, Peterborough, ON: Broadview Press.

Bookchin, M. (1982) *The Ecology of Freedom*, Palo Alto, CA: Cheshire.

Coase, R. H. (1960) 'The problem of social cost', *Journal of Law and Economics* 3: 1–44.

Daly H. (1991) *Steady State Economics, 2nd edn.*, Washington, DC: Island Press.

Daly, H. E. and Cobb, J. B. Jr. (1989) *For the Common Good: Redirecting the Economy Toward Community, the Environment, and a Sustainable Future*, Boston, MA: Beacon Press.

Dryzek, J. (1987) *Rational Ecology: Environment and Political Economy*, Oxford: Basil Blackwell.

Eckersley, R. (1992a) 'Linking the parts to the whole: bioregionalism in context', *Habitat Australia* 20, 1: 34–6.

—— (1992b) *Environmentalism and Political Theory: Toward an Ecocentric Approach*, London: UCL Press.

—— (ed.) (1995) *Markets, the State and the Environment: Towards Integration*, London: Macmillan.

Ekins, P. (1992) 'Towards a progressive market', in P. Ekins and M. Max-Neef (eds), *Real-Life Economics: Understanding Wealth Creation*, London: Routledge.

Elster, J and Moene, K. O. (1989) 'Introduction', in J. Elster and K. O. Moene (eds), *Alternatives to Capitalism*, Cambridge: Cambridge University Press.

Gorz, A. (1980) *Ecology as Politics*, trans. P. Vigderman and J. Cloud, London: Pluto Press.

Habermas, J. (1984) *The Theory of Communicative Action*, Boston, MA: Beacon.

Helm, D. and Pearce, D. (1991) 'Economic policy towards the environment: an overview', in D. Helm (ed.) *Economic Policy Towards the Environment*, Oxford: Blackwell.

Jacobs, M. (1991) *The Green Economy*, London: Pluto Press.

—— (1992) 'The limits to neoclassicism: towards an institutional environmental economics', seminar paper, Institute of Ethics and Public Policy, Monash University, October.

—— (1993) '"Free market environmentalism": a response to Eckersley', *Environmental Politics* 2, 4: 238–41.

Lacey, M. J. (ed.) (1991) *Government and Environmental Politics: Essays on Historical Developments since World War Two*, Washington, DC: The Johns Hopkins University Press; Baltimore, MD, London: Woodrow Wilson Center Press.

Leiss, W. (1990) *Under Technology's Thumb*, Montreal: McGill–Queen's University Press.

McLaughlin, A. (1990) 'Ecology, capitalism, socialism', *Socialism and Democracy* 10: 69–102.

Moran, A., Chisholm, A. and Porter, M. (eds) (1991) *Markets, Resources and the Environment*, North Sidney: Allen & Unwin.

O'Connor, J. (1989) 'Political economy of ecology of socialism and capitalism', *Capitalism, Nature, Socialism* 3, 1: 93–107.

Pearce, D. W. and Turner, R. K. (1990) *Economics of Natural Resources and the Environment*, New York, London: Harvester Wheatsheaf.

Pearce, D., Markandya, A. and Barbier, E. B. (1989) *Blueprint for a Green Economy*, London: Earthscan.

Pepper, D. (1993) *Eco-socialism: From Ecology to Social Justice*, London, New York: Routledge.

Pigou, A. (1920) *The Economics of Welfare*, London: Macmillan.

Ryle, M. (1988) *Ecology and Socialism*, London: Century Hutchinson.

Sale, K. (1985) *Dwellers in the Land: the Bioregional Vision*, San Francisco, CA: Sierra Club Books.

Schmidtz, D. (1993) 'Market failure', *Critical Review* 7, 4: 525–37.

Weston, J. (ed.) (1986) *Red and Green: the New Politics of the Environment*, London: Pluto Press.

Williams, R. (1983) *Towards 2000*, Harmondsworth: Penguin.

Zarsky, L. (1990) 'The green market', *Australian Left Review* 12, December: 12–17.

Index